Molecular Origami

PRECISION SCALE MODELS FROM PAPER

Molecular Origami

PRECISION SCALE MODELS FROM PAPER

Robert M. Hanson

ST. OLAF COLLEGE

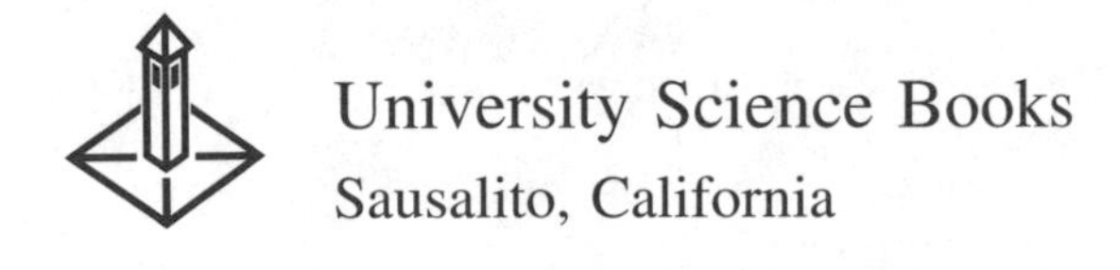

University Science Books
Sausalito, California

University Science Books
55D Gate Five Road
Sausalito, CA 94965
Fax: (415) 332-5393

Production manager: *Steve Peters*
Copy editor: *Kris Landon*
Design consultant: *Robert Ishi*
Compositor: *Integrated Graphics*
Printer and binder: *Braun-Brumfield, Inc.*

This book is printed on acid-free paper.

Library of Congress Cataloging-in-Publication Data

Hanson, Robert M., 1957-
Molecular origami : precision scale models from paper / Robert M. Hanson.
p. cm.
Includes bibliographic references and index.
ISBN 0-935702-30-X
1. Molecules—Models. 2. Models and modelmaking. 3. Paper work. I. Title.
QD461.H263 1995
541.2'2'0228—dc20 94-37451
CIP

Printed in the United States of America
10 9 8 7 6 5 4

Contents

Detailed List of Models

Part 6: One- and Two-Dimensional Species

Preface

The important thing is not to stop questioning. Curiosity has its own reason for existing. One cannot help but be in awe when one contemplates the mysteries of eternity, of life, and of the marvelous structure of reality.

—Albert Einstein

I love to make models. When I was young, I made model cars and model airplanes. The kits I put together, with their seemingly infinite number of highly detailed pieces, were always a challenge. One of the greatest aspects of these kits was that they were *scale* models—all of the pieces were tiny replicas of the real things, all carefully crafted and highly detailed. Something magical happened when the last piece was finally put into place. Suddenly, I held in my hand—or at least in my mind—the "real thing." I could imagine driving that car or flying that airplane. I could turn it over and look at it from all sides. All the pictures in all of the encyclopedias in the world could not equal that experience of making a model myself.

Somewhere along the way I discovered model rocketry. Now the focus was on imagination, design, and performance. No scale models here. Long hours were spent coming up with a new concept, sketching it out as carefully as I could, trying to figure out if it would really fly. I would buy the tubing, balsa wood, and paint. I would order the parts by mail, wait with anticipation for the mail each day, and finally, when the parts arrived, I would spend hours and hours in the basement of our house carefully cutting, gluing, and painting. My dad and I would drive to a field just outside of town, lay out a launching pad, and *Whoosh!* off would go my newest creation, sometimes never to be seen again.

Another passion I had as a child was origami, the Japanese art of paper folding. I learned how one could transform a simple, square sheet of paper into a beautiful model of a bird or other animal by careful, crisp folding. In classical origami no scissors are used, and all of the lines for folding are derived from the original dimensions of the paper.

There is no doubt in my mind that when I was just beginning to learn about science, and especially about chemistry, these early experiences making models played an important role. I chose chemistry above the other sciences because here was a field I could get my hands on. The molecules I read about begged to be made into three-dimensional forms. I was delighted that one of the required components of my first college course in chemistry was a molecular model kit, *Framework Molecular Models*, from Prentice Hall. This particular kit, with its thin, colored plastic tubes and its little metal jacks for connecting them, was especially interesting in that it was truly a scale model kit. I had to measure the tubes for C–C bonds, cutting them just a little longer than the tubes for C–O bonds. This precision work was a valuable exercise, teaching me much about the variations in structure inherent to chemistry.

That molecular model kit still sits on my desk—gathering dust, half hidden by my computer monitor. It served its purpose well—getting my mind involved time and time again, getting me thinking about the mysteries of chemistry. Though it has lost many of its original pieces, I still cherish it, for it represents to me my first foray into the three-dimensional world of chemistry.

Of course, any molecular model you choose to use will teach you something. The models in this book, though, are unique and can teach much more than most models, for they are scale models and data tables all rolled into one. They are snapshots of real life, representations of nuclei caught in the act of jittering and tumbling, caught in a swarm of electrons and unable to escape. Some are large and some are small, some are "regular" and some are irregular. They are as close to the real thing as you will find in terms of detailing the structure of molecules. The scale I have chosen for all of the models in Parts 1 and 6 is 300,000,000:1. At this scale, 1 angstrom is 3 cm (about 1 inch) on the model. In keeping with the trend toward SI units, all distances are indicated in picometer (pm) units. One pm = 1×10^{-12} m = 0.01 angstrom. Thus, the label "H—96—O" on the model for H_2O indicates an internuclear distance between the hydrogen and oxygen nuclei of 96 pm or 0.96 angstroms. Angles are given in degrees. In Parts 2–5, some of the models are scaled down by 20–60% so as to fit on the page.

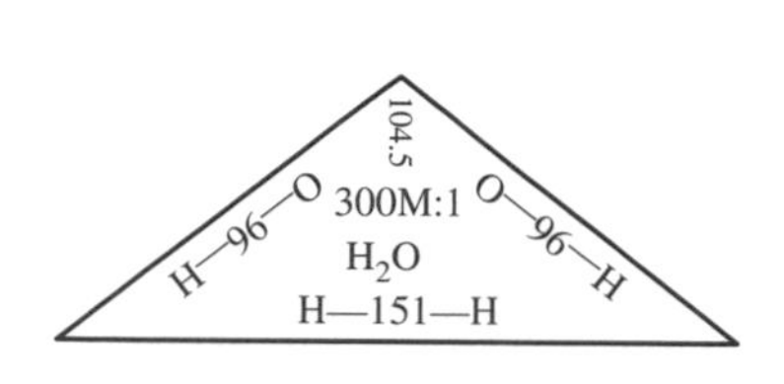

The distances and angles indicated on the models have much to say, for in them lie the secrets of molecular bonding. Why is NH_4^+ so much smaller than BH_4^-? Why is SF_4 so totally different from SiF_4? What happens to the structure of PF_3 when it is oxidized to form POF_3?

No, you won't be able to "launch" these models into reactions the way you might launch a model rocket into flight. They are static. But the scientific model you build *in your mind* to explain the data presented by these paper models should serve you well when you go into the laboratory and launch your own "model" reactions.

I recognize that the method of folding and cutting presented in *Molecular Origami* does not even come close to doing justice to classical origami, with its focus on idealized forms. (Most of the basic patterns in this book actually can be folded from a square or rectangular sheet of paper in the classical sense, but even then scissors and tape are required.) In *Molecular Origami*, the beauty is not so much in the paper molecular models as it is in the molecules themselves, which somehow get "folded" by nature without any help from us. Nature, after all, is the true origami master.

Molecular Origami: Precision Scale Models from Paper is designed to be a resource for students and teachers of chemistry who want to have models of real molecules and ions that they can hold in their hands, pass around a group, examine carefully, and generally have fun with. What you are holding, basically, is a set of data tables and molecular models all rolled into one. All of the information provided is from real data, the results of real experiments; references to the original sources and methods used are included.

Part 1 focuses on two of the most common three-dimensional molecular shapes: the trigonal pyramid and the tetrahedron. Part 2 adds four more shapes, up through the octahedron. Part 3 touches on six more shapes that are rarely modeled. In Part 4 are a few molecules and ions that involve more than one "central atom." Part 5 presents quartz, an example of a "network solid." Finally, in Part 6 are 76 one- and two-dimensional models which require no folding. These models are presented on a scale identical to the models of Part 1 so that you can directly compare distances and angles among all of these species. I have used this section in conjunction with models of Part 1 to introduce Lewis structures to my students, where the goal is to explain the overall geometry and to see the variety of sizes and shapes. What angles are really observed for "bent" molecules such as H_2O, and OF_2? Then, later on, I also provide models from Part 2 when I'm introducing VSEPR theory and some of the more subtle ideas of bonding.

My goal in writing this book is to stimulate your curiosity about chemistry and the structure of matter. You will find yourself asking questions. Why is this structure so large or small? Why is this molecule puckered and this one not? Why are the angles the way they are? Why is this bond so long and this other so short? What sort of trends can I find in these data? These are the sorts of questions that chemists love to ask. The answers invariably bring in ideas of electrons, protons, and bonding. To get you started, I've supplied with most of the structures in Part 1 a few "questions to think about." I'm sure you will come up with many more. How you answer these questions, of course, will depend upon your own particular perspective and background—there is never one "right" answer.

A discussion of the questions is included which uses molecular orbital theory as a basis, presenting the elements of this theory that are necessary for understanding the arguments involved. Throughout the discussion the focus is on a comparison of two models at a time. (After all, one has but two hands!) This section can be read and appreciated even without taking the time to actually make any models, but the comparisons are more fun if models are in hand. At the end of the Discussion section is a list of nine trends that I noticed in the data. Some of them you may recognize, and some may be new to you. Also at the end of this section is a summary of ten basic ideas from molecular orbital theory used throughout the discussion.

You will find the index particularly useful for making meaningful comparisons. Every comparison mentioned in the discussion is indexed, and each index entry points to both the data and the discussion. A class exercise, for example, might be to compare the structures listed under the heading "ammonia, NH_3, comparison with . . . ."

However you use it, I hope you have as much fun working with *Molecular Origami* as I have had putting it together. Feel free to contact me with any questions or comments you may have.

Robert M. Hanson
Department of Chemistry
St. Olaf College
Northfield, MN 55057 USA
E-Mail: hansonr@stolaf.edu

Acknowledgments

This book is the culmination of many influences. Above all, I owe thanks to Ellen Heeren, recently retired from Arapahoe High School in Littleton, Colorado, who inspired my curiosity about the wonder of nature. Special thanks are due to Marian (Hawkinson) Ano for her introducing me to the art of classical origami. Thanks also go to Sara Bergman, Karl Nelsen, Jim Baron, and Chris Rasmussen, who, as St. Olaf students, helped immensely with the research underlying this book. Finally, I wish to thank Debbie, Ira, and Seth for their patience, especially during the wee hours of the night when much of this work was accomplished.

—R. M. H.

Molecular Origami

PRECISION SCALE MODELS FROM PAPER

Introduction

Because many students find difficulty in appreciating three-dimensional structures from two-dimensional illustrations, the examination of, and preferably also the construction of, models should play a large part in the study of these subjects.

—A. F. Wells, 1984

How to Fold the Models

The patterns in this book are all labeled for easy folding, although it is true that some models are more challenging than others. The model pages are perforated for removal; if you prefer to photocopy the page instead feel free to do so (within the bounds of copyright law). Standard 20-lb. photocopy paper, which is somewhat lighter than the paper in this book, is perfect for folding. Then get out your scissors and tape and get to work! The rules are simple and typical of origami:

Solid lines are "mountain" folds, which fold away from you.
Dashed lines are "valley" folds, which fold toward you.

In addition, the shaded regions can be either removed or simply folded back and tucked into the hidden recesses of the model. Removing them may be quicker, but I definitely recommend leaving them in. They actually make folding easier and add to the strength of the model. The figure below shows the basic sequence, in this case for methane, CH_4. In the first step, lines are creased. Next, shaded regions are brought together and folded behind faces. Finally, the three outer faces (one is hidden) are folded back and secured in place. The points marked with a ● all end up at the same place in the end. A generous application of cellophane tape around all of the edges is advisable.

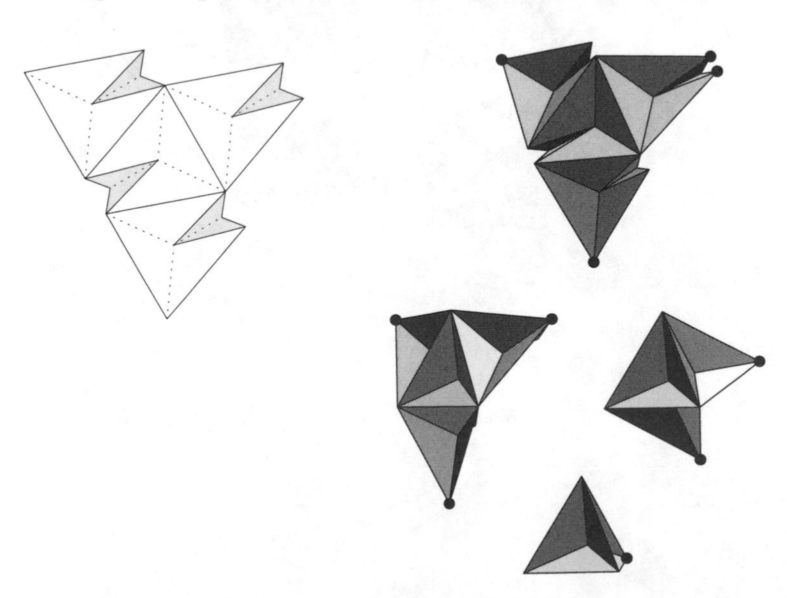

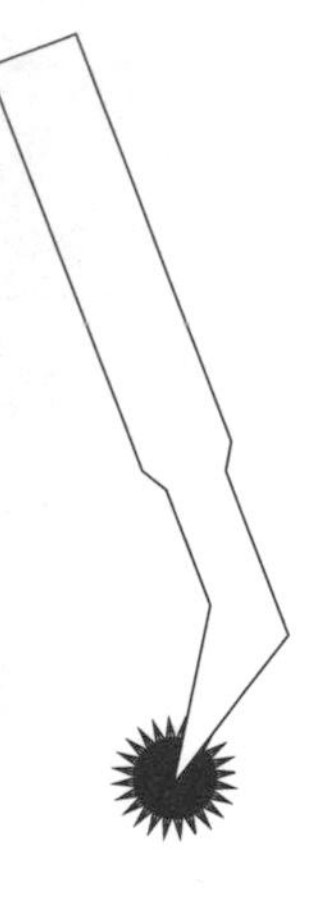

Two basic methods of folding are suggested. The first requires an inexpensive fabric tracing wheel (illustrated on the left); the second employs a credit card or library card.

Using a tracing wheel:

- Place a piece of cardboard under the pattern to be folded.
- Press gently as you wheel along the fold line from the center out. Rulers are unnecessary.
- Turn over the pattern to ensure that all lines are perforated.
- Cut out the pattern (if it isn't already) and fold as indicated.

Using a credit card:

- Cut the pattern out along its outer edges.
- Lay the card (the thinner the better) along the top edge of a line to be folded.
- Fold up and away from you, drawing along the fold with the edge of your fingernail.
- When folding the solid lines, turn the model over and place it on a blank sheet of white paper. This allows you to see the folding line clearly through the paper and still fold toward you over the card.

In addition, for some of the more complicated structures, two or more units must be put together. There are four basic ways this can be accomplished:

- **Edge-to-edge.** For example, in B_2H_6 (page 143). This is easy.

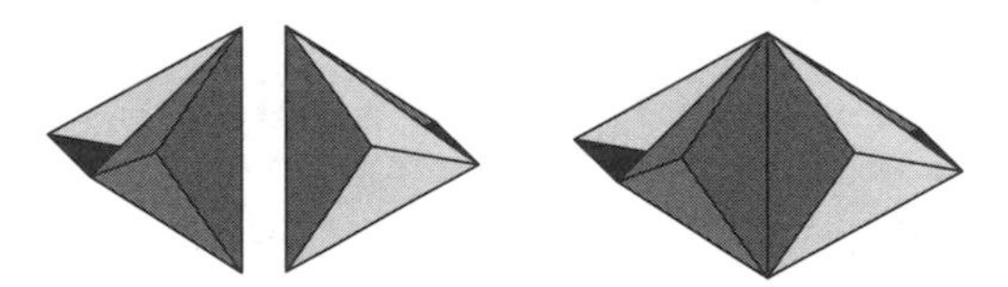

- **Face-to-face.** This is a bit harder and is used for $Fe_2(CO)_9$ (page 145).

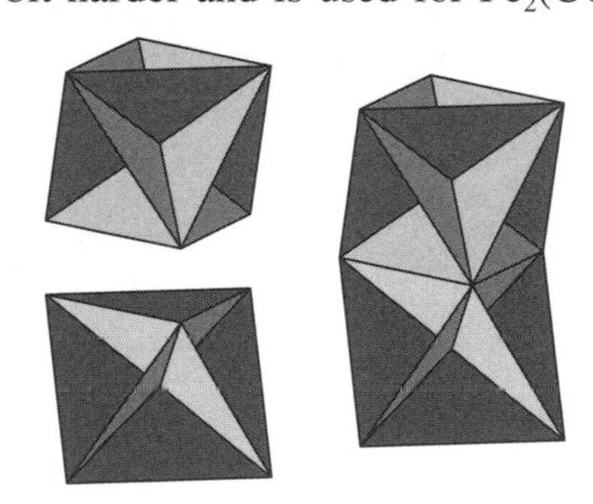

- **Corner-to-corner.** This requires "tabs" which are usually overlapped to provide the necessary alignment. The quartz model (page 161) will require this method.

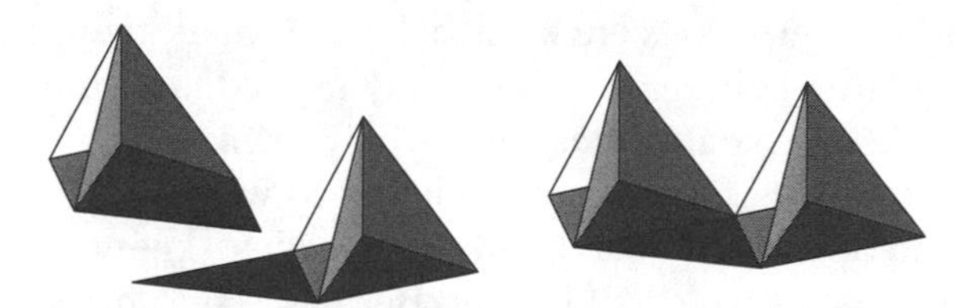

- **Center-to-corner.** (This is what I call *splicing* or *chaining*.) Examples include ethane (page 49) and methylamine (page 53). Splicing involves making a single cut in each subunit and fitting them into each other like this:

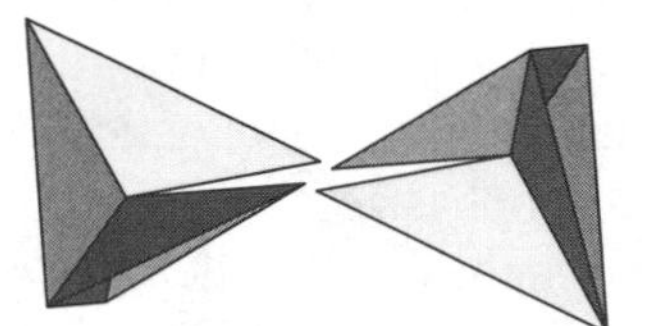

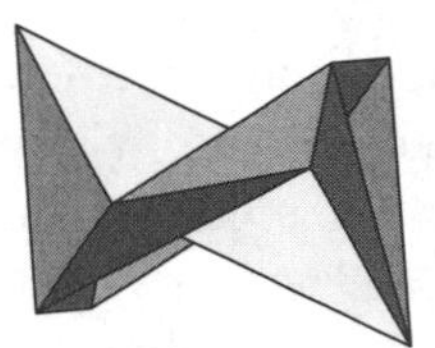

Molecular Structure and Bonding

For anyone trying to visualize the structure of matter, good molecular models are essential. Balloon models, plastic or wood "ball-and-stick" models, expensive "space-filling" models, and computer graphic models are all available and have their individual advantages and disadvantages. They range from the very cheap to the extremely expensive; from concrete to intangible, from flexible to rigid, from crude to precise. Nonetheless, they all have one thing in common: They all are based on chemical *bonding*. What makes the models in this book unique is that they are are based on *structure*, not bonding.

Bonding and structure are not one and the same. Molecules are composed of atoms, and atoms are composed of nuclei and electrons. When chemists speak of *molecular structure*, they are referring to the detailed three-dimensional positions of the nuclei in a molecule. Those positions are exactly what these folded paper models will give you.

When chemists speak of *bonding* in a molecule, on the other hand, they are referring to the arrangements in space of the valence (outermost) electrons. Chemists distinguish two types of electrons in molecules: *core* and *valence*. Core electrons are held extremely tightly by nuclei and are never shared with other nuclei. Valence electrons are held more loosely and, thus, can be used for connections with other nuclei—that is, for bonding.

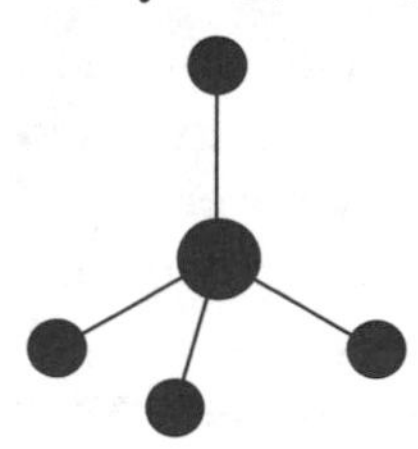

So bonding has to do with valence electrons; and structure has to do with nuclei and their associated core electrons. Typical ball-and-stick models are structural in the sense that the balls are in certain places. The problem, as I see it, is with those sticks. Are they bonds, or are they just means of placing the balls? Do they indicate the locations of the electrons or "electron density," or are they just supports? The fact is, it is often just too tempting to make them into more than they really are. Basically, these models introduce us to only the crudest idea of bonding that was popularized about 50–60 years ago. More recent experimentation (mostly "photoelectron spectroscopy" in the 1970s) has shown unequivocally that even in molecules as simple as CH_4, NH_3, and H_2O the electrons are *not* localized. Rather, they are spread throughout the molecule, each electron helping in its own way to hold the whole molecule together.

Those sticks are like the balance statement from a bank telling you how much money you have in what account but giving you no clue as to what the bank is doing with your money. As a consumer, perhaps that is all you need to know. At the entry level of chemistry, perhaps that is all we should teach. But wise consumers, economists, capitalists, and many others realize that banks do more than just stash their money. Anyone involved in any way with molecules should be aware that models that depict electron positions as sticks will only go so far.

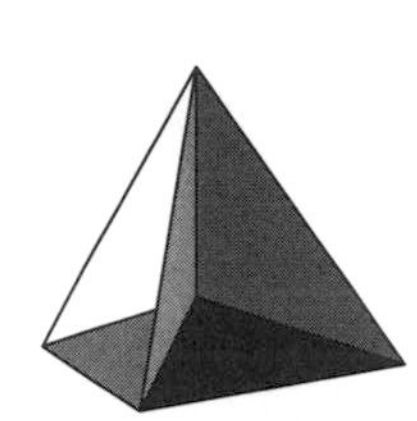

My goal in presenting these models is quite simple: to provide you with a different kind of structural model, a sort *lacking implicit assumptions about bonding*. With these models the molecular structure is completely defined. Instead of "atoms" we have "nuclear positions;" instead of "bonds" we have "internuclear distances." Some of these distances might be interpreted as bonds and some not. It is up to you. Your choice will emphasize your inclination toward localized or delocalized bonding. I happen to like to emphasize the delocalized "whole-system" approach to bonding, but you may want to emphasize the idea that some atoms are connected and some are not, and that those connections are pairs of electrons. If you want to draw in the lines and call them pairs of electrons, you can. If you want to explain the structures using localized theories such as "VSEPR," you can. It's your choice. What follows is a short synopsis of molecular orbital theory.

Molecular Orbital Theory in a Nutshell

In order to discuss the structures in this book in terms of bonding, we are going to use a model of molecular bonding that is based on *orbitals*—distinct places with distinct energies where electrons can be expected to be found. We'll always be interested in dividing the electrons into two sets, *bonding* and *nonbonding*.

Bonding electrons can be expected to be found primarily in orbitals that are shared extensively among atoms; nonbonding electrons, on the other hand, can be expected to be found in orbitals mostly localized on specific atoms. Thus, we might talk about N–H "bonds" and N "lone pairs."

It is important to keep in mind that bonding electrons aren't usually shared by just two atoms, and nonbonding electrons don't necessarily have to be localized. After all, electrons in molecules, if considered to be particles, are traveling at anywhere from 1% to 99% of the speed of light. How could they possibly be expected to confined to such small areas? More precisely, it is the electron *density* which is localized. The electrons themselves are exchanged among atoms at an incredible rate. In any case, it is convenient to talk about electrons as though they were localized, at least to get started.

The real key to the terms "bonding" and "nonbonding" is energy. Each electron—each orbital—is thought in this simple model to have a distinct energy in relation to its neighbors. In atoms, electrons closer to the nucleus (on average) generally have lower energy than electrons further from the nucleus. Atomic electrons appear to be localized into spherical "shells." Molecules appear to result from sharing only of the electrons in the outermost atomic shells. These electrons are called the *valence* electrons, and the atomic orbitals holding them are called the *valence orbitals*.

Within an atomic shell, the order of increasing energy is *s*, *p*, *d*. Thus, electrons in *s*-type orbitals are (on average) lower in energy (more stable, harder to remove, less likely to be involved in bonding) than electrons in *p*-type orbitals. This difference in energy is due to *shielding*. *s* electrons spend more time closer to the nucleus than do *p* electrons. Therefore, *s* electrons experience a stronger nuclear attraction and are held more tightly than *p* electrons. In effect, the *s* electrons' presence shields the *p* electrons from the nucleus by canceling out a bit of nuclear charge. This shielding is not totally effective, because the *s* electrons are only *on average* closer to the nucleus than the *p* electrons. Nonetheless, the effect of shielding on making *p* electrons higher in energy than *s* electrons is significant.

Most simple molecules such as NH_3 can be thought of as a central atom surrounded by a small number of outer atoms. The bottom line in molecular bonding is that there is only so much central atom *s* and *p* orbital "character" to go around. Nonbonding electrons on the central atom (lone pairs) can be expected to take more than their fair share of the central atom's *s* character, because those electrons have the most need to be stabilized by the central atom itself.

One of the keys to understanding molecular structure is in understanding that *s* character is "spherical" while *p* character is pointed in specific directions. In fact, it would be safe to say that the only reason there is any "structure" to molecules is because *p* orbitals have direction. If the bonding in NH_3 were due solely to *s* orbitals, there would be little reason for the H atoms to be in any particular position, and the molecule would be a floppy set of four nuclei with no well-defined shape.

A second key to understanding the structure (as well as the reactivity, by the way) of small molecules may at first seem a bit ridiculous: Focus on the central atom's *nonbonding* electrons. They are the ones that benefit most from stabilization, either by taking on central atom *s* character (leading to structural definition) or by delocalizing onto other molecules (leading to reaction.)

So, in a nutshell, we have protons, neutrons, and electrons, of which we can pretty much ignore the neutrons. Electrons are attracted to protons and repelled by other electrons. Chemists focus on the electrons when talking about bonding and reactivity, because electrons are "where the action is." Electrons are going so fast, in fact, that about the only way to think of them is with probability. The regions of space in which we can expect electrons to be found are called orbitals.

Some orbitals, when they contain electrons, contribute to the holding together of the molecule. We call these bonding orbitals. Some orbitals contribute to the breaking apart of the molecule (antibonding orbitals), and some are not particularly important for keeping a molecule together (nonbonding orbitals). It turns out that the nonbonding electrons, though not really very important for bonding, are essential in defining the structure and reactivity of molecules. Finally, some molecular orbitals extend over many atoms (delocalized orbitals), and some pretty much reside on specific atoms (lone pairs and core orbitals). Principally when we are talking about structure we are interested only in the valence electrons—just the ones of highest energy, the ones most likely to be shared among atoms.

We can think of all molecular orbitals as having certain proportions of *s*, *p*, and sometimes even *d* character. The idea of character comes from our use of atomic orbitals as building blocks for molecular orbitals. It is this character that gives a molecular orbital shape and energy. It is this character that people are talking about when they refer to "sp^3 hybridization," and it is this character, especially for the central atom, which defines the shape of a molecule. *s* character is spread out and spherical, while *p* character is more directed.

As a general rule of thumb, isolated molecular systems tend toward the shape that gives them the lowest possible overall energy. Thus, energy is the real key to structure. Adding *s* character contributes to the lowering of electronic energy (possibly at the expense of other electrons), while adding *p* character generally leads to a higher energy state.

It is this tension in orbital character, between the lower-energy, not-so-good-for-bonding *s* character and the higher-energy better-for-bonding *p* character, that makes for the variety of molecular structures found in nature.

PART 1

Basic Shapes, Basic Ideas

Molecules and ions come in many, many shapes. Some are one-dimensional, such as carbon dioxide, CO_2. Some are two-dimensional, like H_2O or BF_3. The vast majority, though, are three-dimensional. The two basic three-dimensional molecular shapes are the **trigonal pyramid** and the **tetrahedron**:

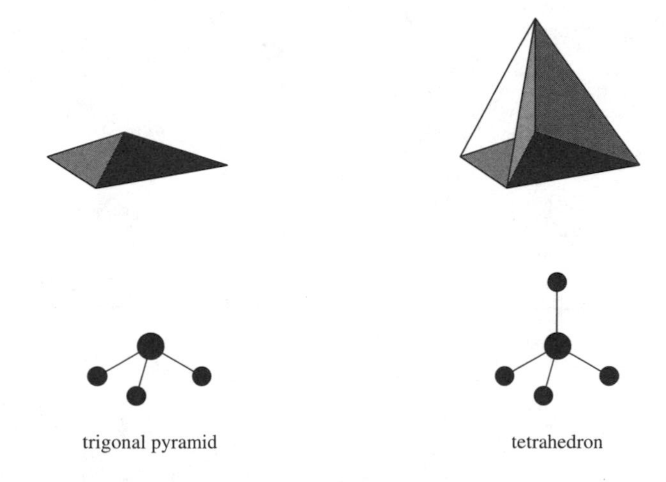

The trigonal pyramid consists of four atoms, with one "central" atom situated above the plane of the other three. In the tetrahedron, an extra atom is found above the other four, making a total of five atoms—one central and four surrounding atoms.

Now, there are many more molecular shapes, but these two provide good starting points. Comparing them, we can learn much about structure and bonding. Why are some of the pyramids steep and some shallow? Why are some tetrahedra "regular" and others not? Is it possible to meaningfully relate pyramid shapes to tetrahedron shapes? In most cases, there are simple answers for these questions. Considering them takes us on a journey into just what exactly a chemical "bond" really is. More than anything, comparing structures gives us an appreciation for the beautiful variety of nature.

In this part of *Molecular Origami*, you'll find patterns for 11 trigonal pyramids and 21 tetrahedra. The idea of having so many of these two shapes is to help you learn by comparison. Questions are given to get you thinking about why the structures might be the way they are. Answers—at least my perspective on them—are in the *Discussion* section. You're certainly welcome to ignore the questions and just have fun making the models, or, alternatively, not make the models and just use the patterns as fancy "data tables" to help you answer the questions. Either way, you'll probably find yourself asking your own questions and designing your own answers reflecting your own particular background.

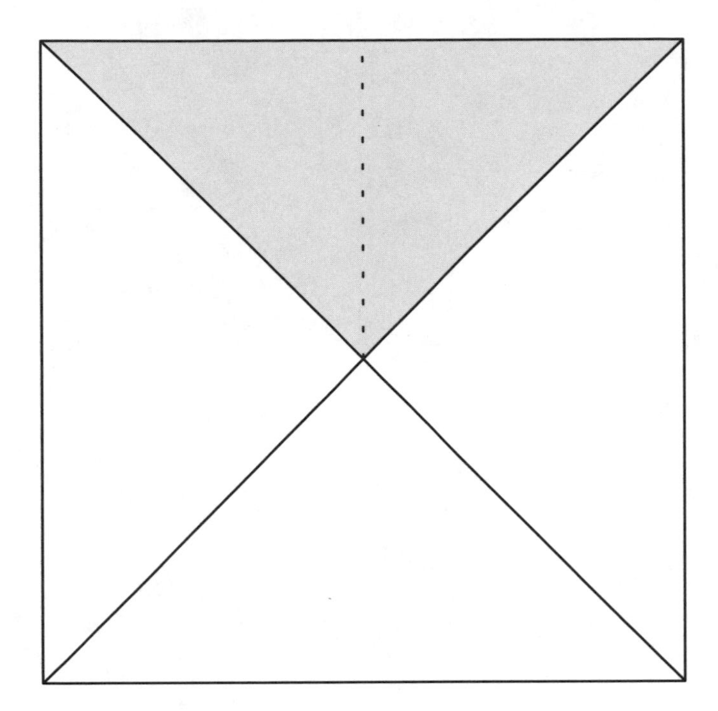

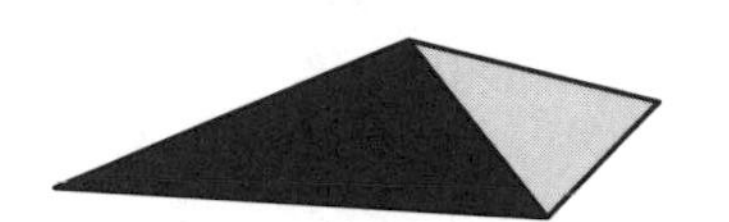

Trigonal pyramidal molecules consist of four atoms in the shape of a triangular-based pyramid. The 'E' in AX_3E stands for *lone electron pair*. All of these molecules have an extra lone pair of electrons primarily associated with the central atom. Calculations indicate that this pair of electrons is directed away from the three outer atoms, thus leading to the observation that AX_3E molecules are nonplanar.

An idealized pyramidal model can be made from any square sheet of paper without use of a pattern simply by folding along the two diagonals and making a vertical crease from the center. As the top two corners are drawn together to make a single X atom, the model will assume three-dimensional form.

The models in this section have a second unit attached just for aesthetics. The second unit ensures that both the top and the bottom of the model bear distance and angle information. In addition, having a second unit means that all shaded regions are completely hidden.

The angles X–A–X in these models vary from 93.8° in PH_3 (page 19) to 112.2° in NH_2CH_3 (page 17). Part of the lesson here is why some of these angles are small and some are large. There are no "right" answers here, merely suggestions that relate to one theoretical model or another. Possible explanations based on molecular orbital theory are given in the *Discussion* section.

ammonia

shape: trigonal pyramidal units: pm scale: 300,000,000:1

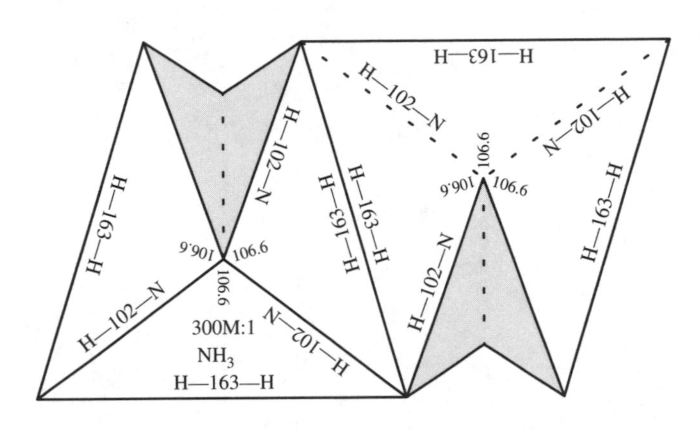

Questions to think about:

(a) Why isn't this molecule flat?

(b) Both CH_4 and NH_3 consist of 10 protons, 10 electrons, and several neutrons. Compare this structure with that of CH_4 (page 31). How are they similar? How are they different?

(c) Imagine "magically" turning CH_4 into NH_3 by moving one of the protons we call "H" into the carbon nucleus to form "N." How are the structural differences between CH_4 and NH_3 consistent with this transformation, given that protons attract electrons?

(d) Imagine magically turning NH_3 into H_2O the same way. What do you predict for the structure of the water molecule?

(e) Ammonia and water react as follows to form the ammonium ion (NH_4^+) and hydroxide (OH^-):

$$NH_3 + H_2O \longrightarrow NH_4^+ + OH^-$$

What do you predict (shape, angles, and distances) for the structure of NH_4^+ (page 35)?

(f) Ammonia and boron trifluoride (BF_3, which is flat) react to form $BF_3{\cdot}NH_3$. What do you predict for the structure of $BF_3{\cdot}NH_3$ (page 41)?

nitrogen trifluoride

shape: trigonal pyramidal   units: pm   scale: 300,000,000:1

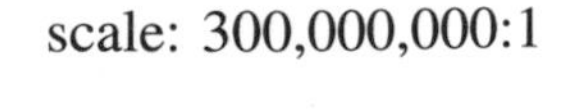

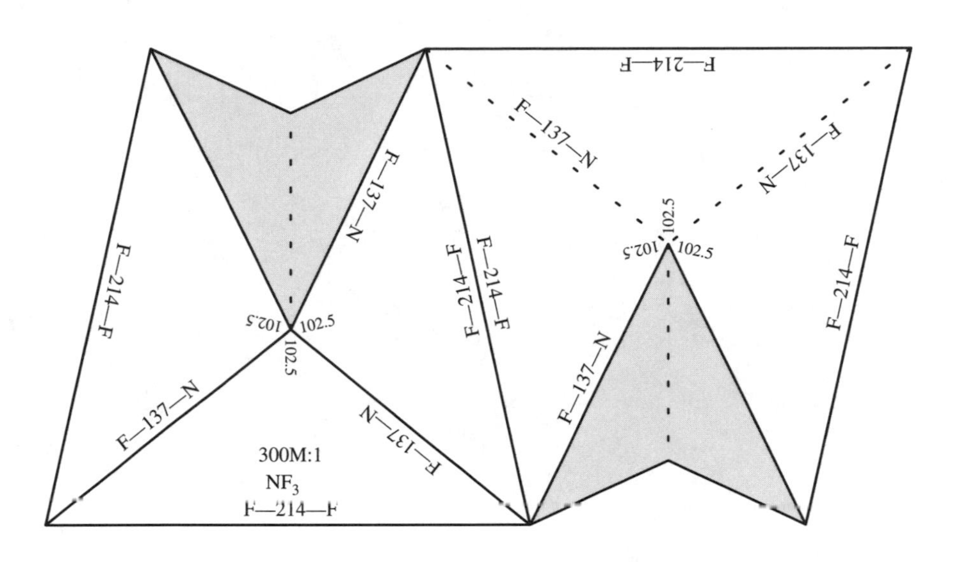

Questions to think about:

(a) How does the structure of NF_3 compare to that of NH_3 (page 11)?

(b) Compare this structure to that of BF_3, shown to the same scale on the right. What are the differences? Explain them.

(c) What would you predict for the distances and angles in CHF_3 (page 57)?

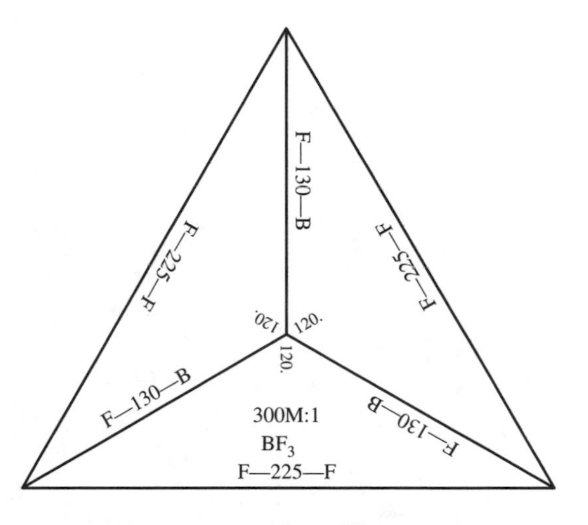

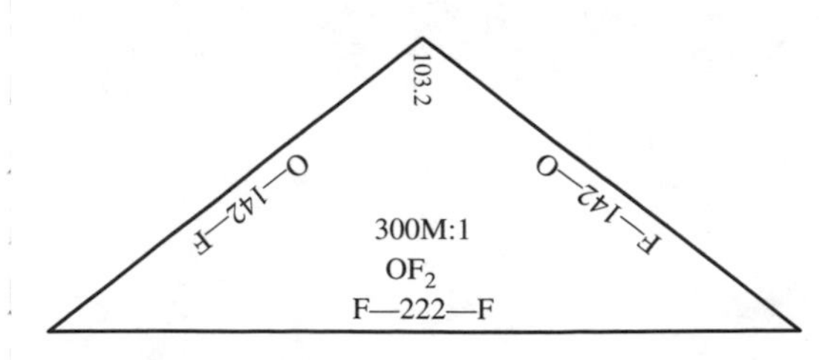

(d) In CF_4 (page 33), the C–F distances are all 132 pm and the angles are all 109.5°. The structure of OF_2 is shown on the left. Is there a trend here as there is for the series CH_4, NH_3, H_2O?

nitrogen trichloride

shape: trigonal pyramidal   units: pm   scale: 300,000,000:1

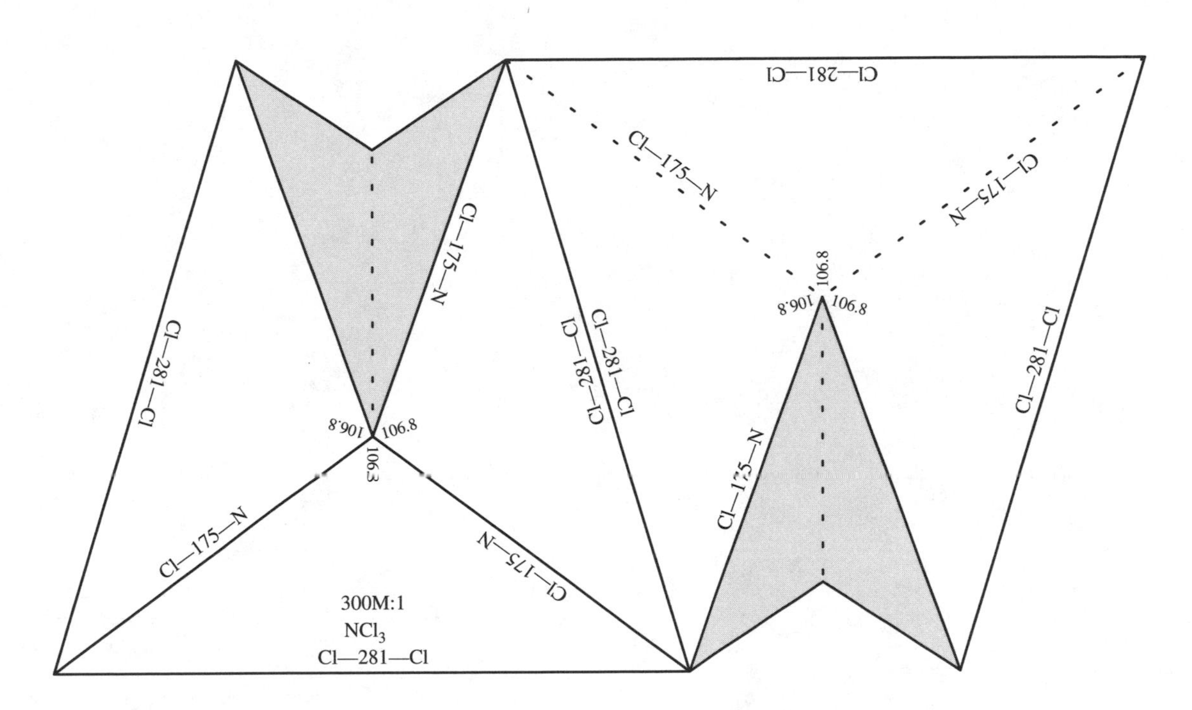

Questions to think about:

(a) What are the main differences between this structure and that of NH_3 (page 11) and NF_3 (page 13)?

(b) Why are the distances in NCl_3 so much larger than those in NF_3?

(c) Why do you think the angles in NCl_3 are almost the same as those in NH_3?

(d) What do you predict for the structure of $CHCl_3$ (page 59)?

(e) Shown to the right is the structure of OCl_2. Compare the structure of OCl_2 to that of NCl_3 and predict the distances in CCl_4.

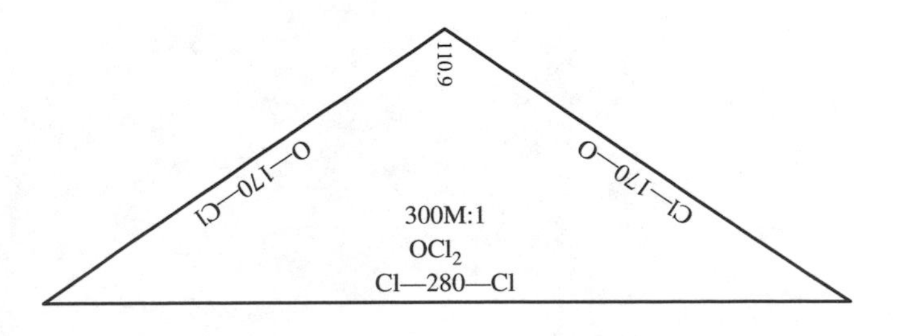

methylamine (N) — NH_2CH_3
difluoroamine — NHF_2

shape: trigonal pyramidal | units: pm | scale: 300,000,000:1

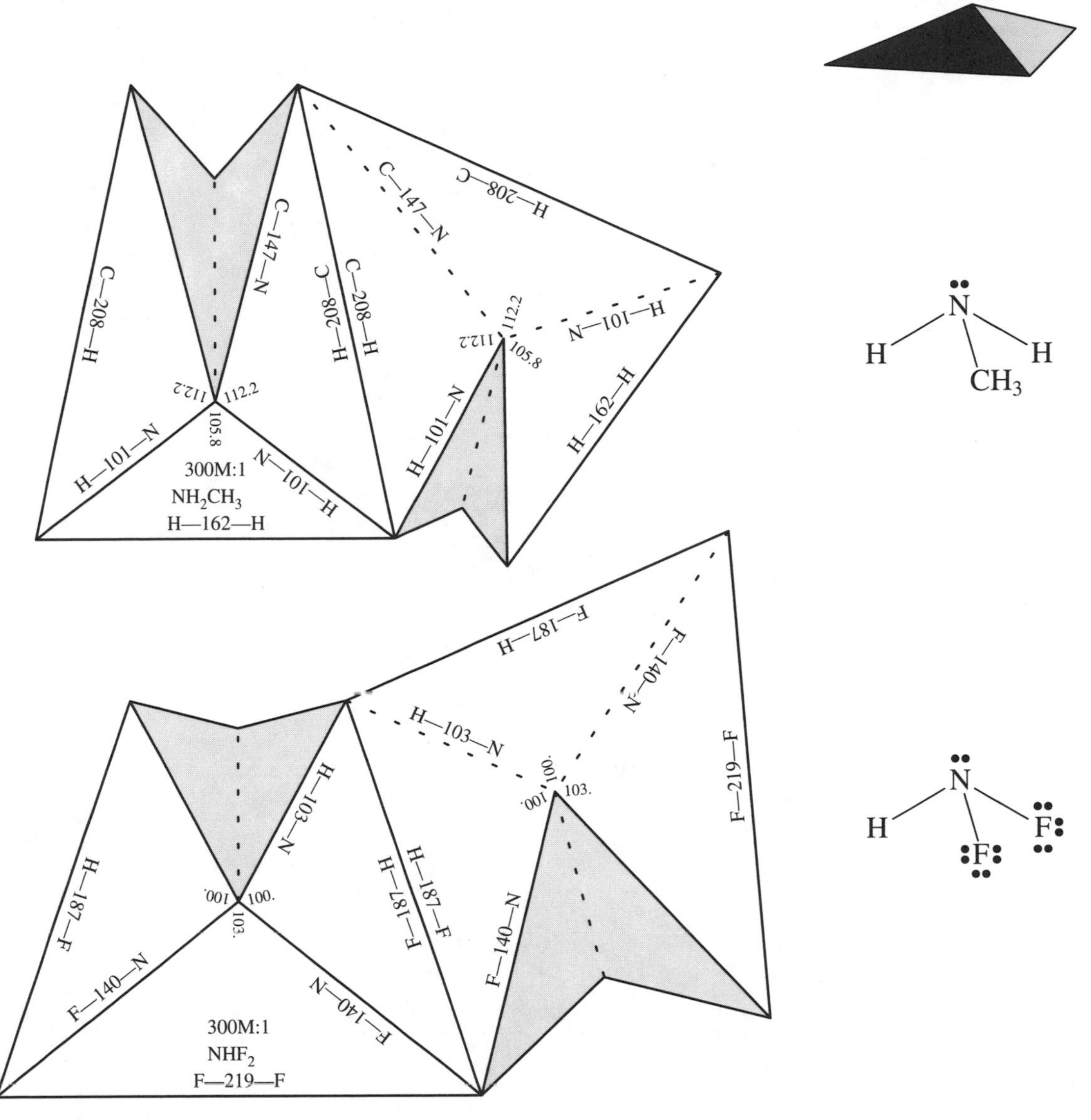

Note: The other half of methylamine is given on page 53.

Questions to think about:

(a) How does the structure of NH_2CH_3 compare to that of NH_3 (page 11)? How do you explain the differences?

(b) Compare the structure of NHF_2 to that of OF_2 (see page 13). How do you explain the differences in X–F distance?

(c) Imagine pulling one of the protons in the N out of each of these molecules to form CH_3CH_3 (page 49) and CH_2F_2. What do you predict for the structures of CH_3CH_3 and CH_2F_2?

phosphine

shape: trigonal pyramidal | units: pm | scale: 300,000,000:1

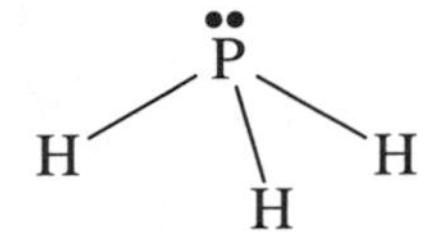

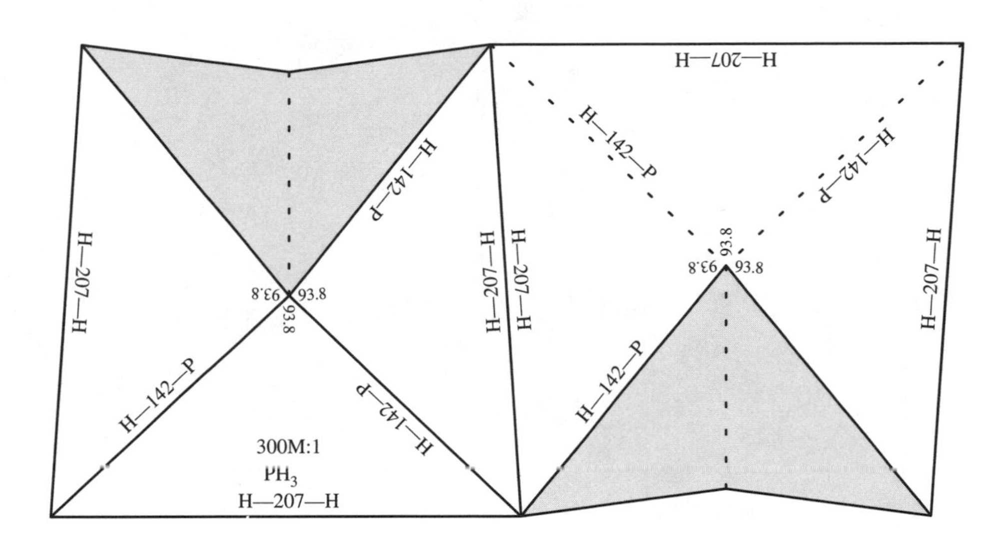

Questions to think about:

(a) Phosphorus is just below nitrogen on the periodic table. How does the structure of PH_3 compare in shape, distance, and angle to that of NH_3?

(b) What do angles near 90° mean in molecular structure?

(c) Shown to the right is the structure of H_2S shown to the same scale. How do you explain the differences in structure between H_2S and PH_3?

(d) What would you predict for the shape, distances, and angles of SiH_4 (page 61)?

(e) Why might PH_3 be less basic than NH_3?

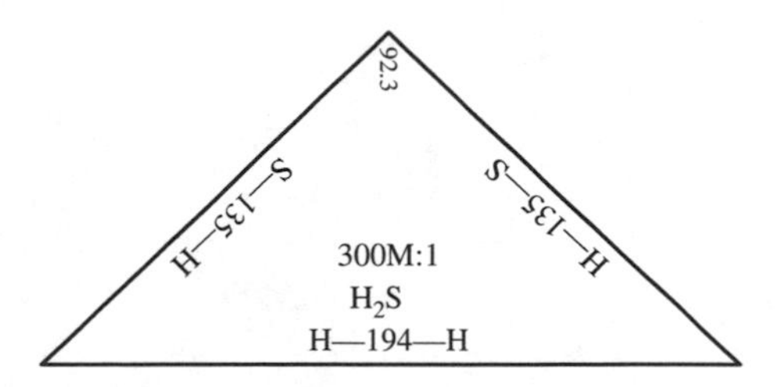

phosphorus trifluoride

shape: trigonal pyramidal   units: pm   scale: 300,000,000:1

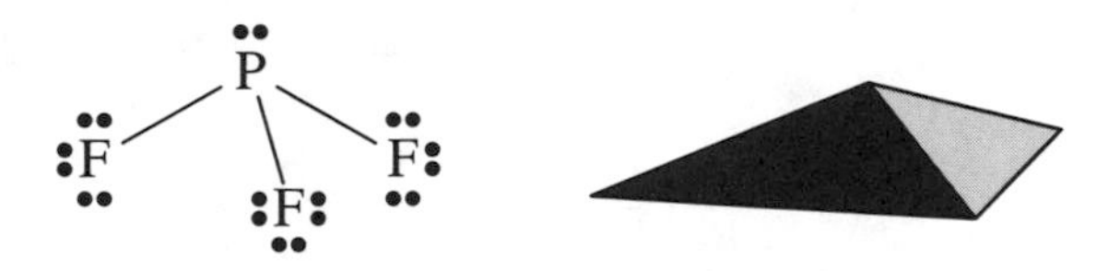

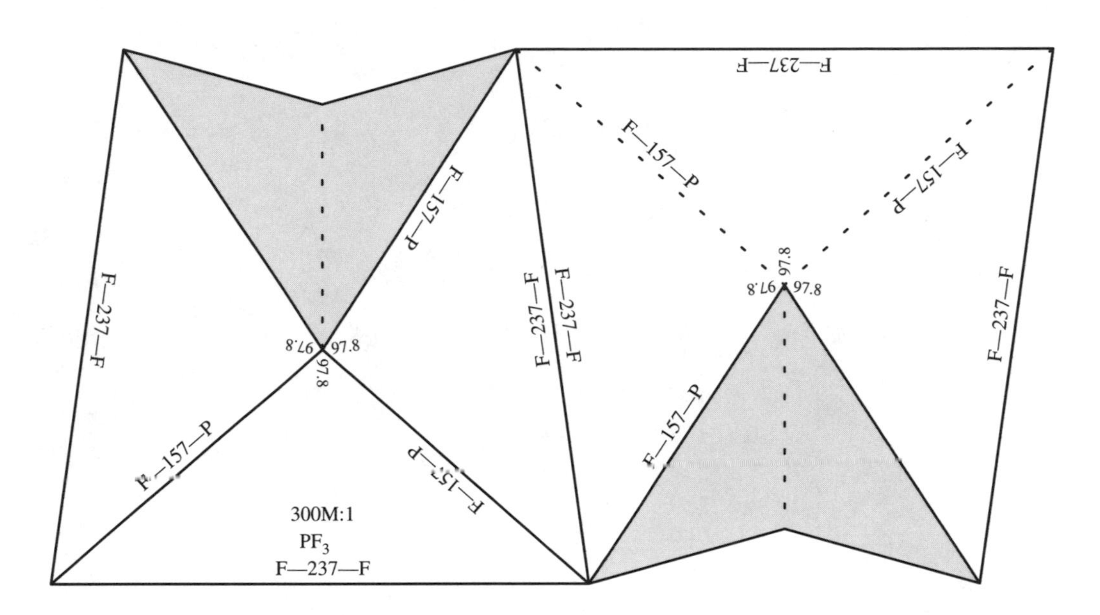

Questions to think about:

(a) NF_3 has smaller angles than NH_3. Why might the angles in PF_3 be *larger* than the angles in PH_3 (page 19)?

(b) Why might the angles in PF_3 be smaller than the angles in NF_3 (page 13)?

(c) Compare this structure to that of its oxidation product, POF_3 (page 65). How do you explain the differences?

(d) What is the connection between "oxidation" as discussed here and the sort of oxidation when Fe^{2+} becomes Fe^{3+}?

(e) What do you predict for the structure of $SiHF_3$?

(f) Shown on the right is the structure of SF_2, an extremely unstable molecule. Comparing these two structures with that of SiF_4 (page 63), what trends in angle and distance do you see for fluorides across the second row? Are these the same trends seen in the series CF_4 (page 33), NF_3 (page 13), OF_2 (page 183)?

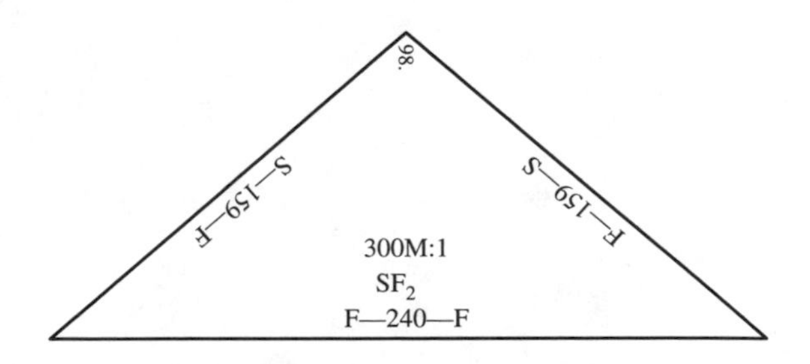

phosphorus trichloride

shape: trigonal pyramidal   units: pm   scale: 300,000,000:1

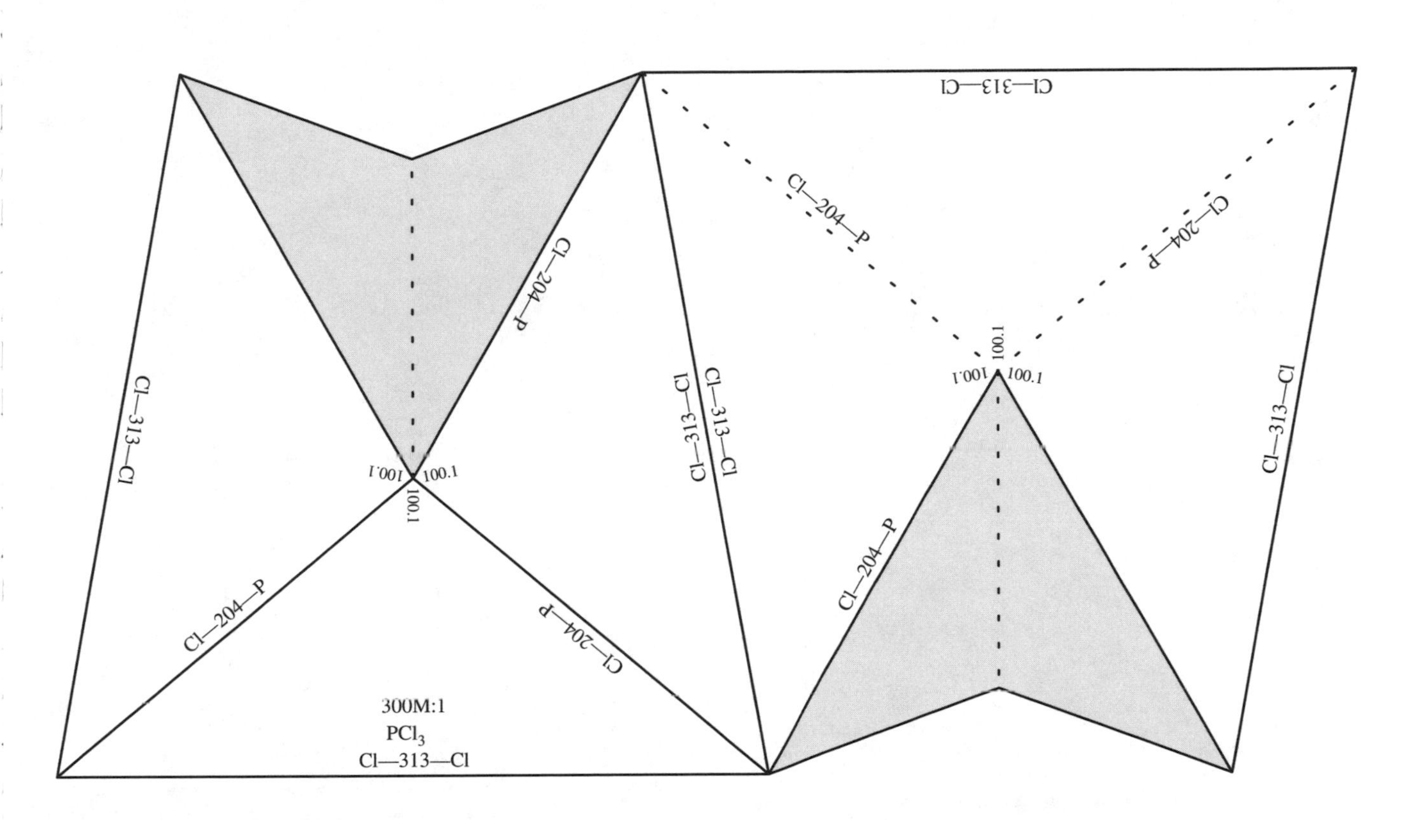

Questions to think about:

(a) Compare the distances in PCl_3 to those in PF_3 (P–F 157 pm) and PBr_3 (P–Br 220 pm). Are the distances in PCl_3 reasonable?

(b) How does this structure compare to that of NCl_3 (page 15)?

(c) Shown to the right is the structure of SCl_2. What does a comparison of the structures of $SiCl_4$ (Si–Cl 201 pm), PCl_3, and SCl_2 teach us?

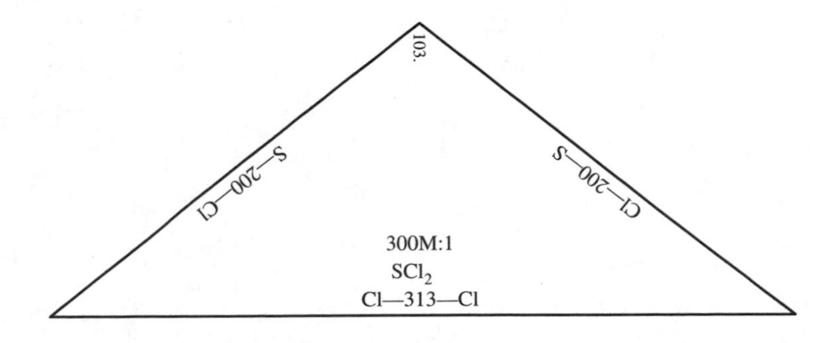

(d) What would you expect for the structure of $POCl_3$?

(e) What would you expect for the structure of $SiHCl_3$?

difluorophosphine

shape: trigonal pyramidal | units: pm | scale: 300,000,000:1

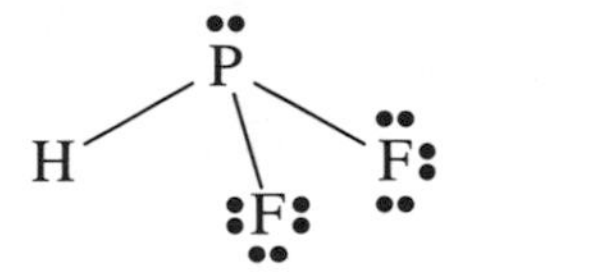

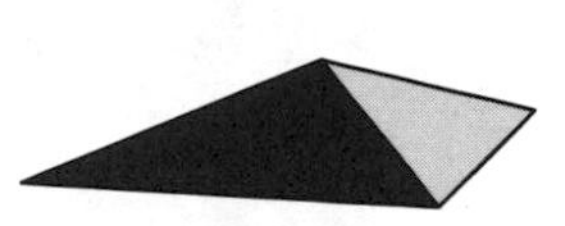

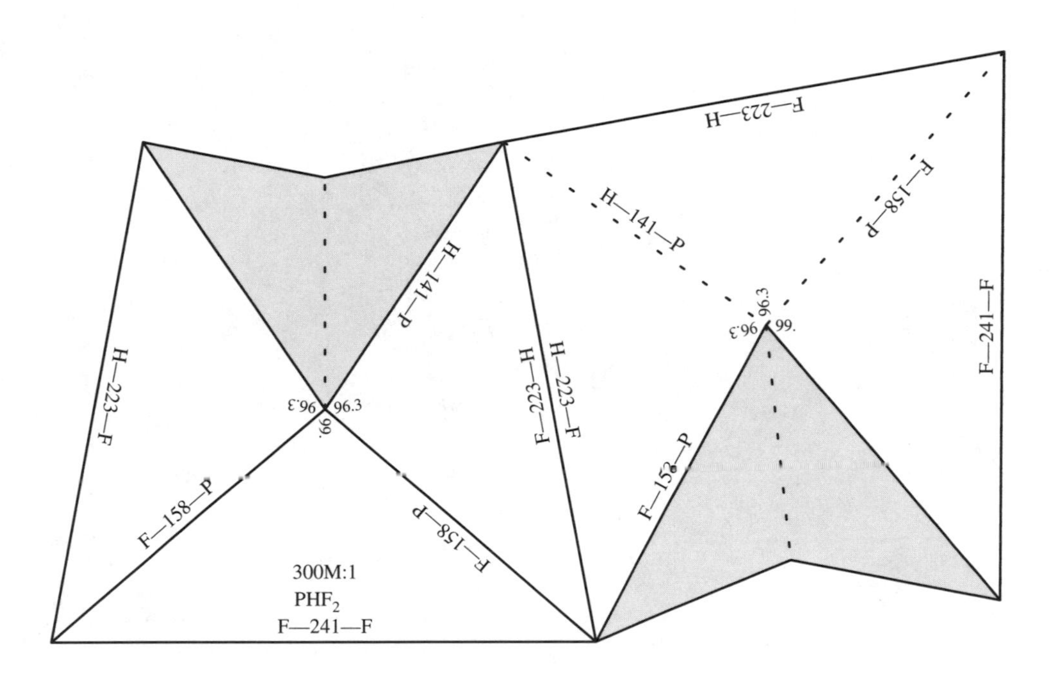

Questions to think about:

(a) How is PHF_2 related to SF_2 in terms of protons and electrons?

(b) Shown to the right is the structure of SF_2. Do the very slight differences in structure of PHF_2 and SF_2 make sense?

(c) Imagine adding an oxygen atom to the phosphorus to make $HPOF_2$. What would you expect its structure to be like?

(d) Phosphorus is just below nitrogen on the periodic table. How are the structures of NHF_2 (page 17) and PHF_2 related?

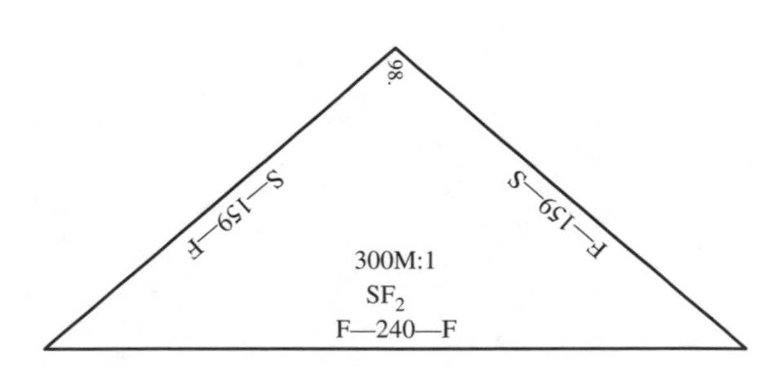

iodate ion (in NH_4IO_3) — IO_3^-
xenon trioxide — XeO_3

shape: trigonal pyramidal   units: pm   scale: 300,000,000:1

300M:1
IO_3^-
O—284—O

300M:1
XeO_3
O—275—O

Questions to think about:

(a) Why might these two structures be so similar?

(b) Why might the distances in IO_3^- be larger than those in XeO_3?

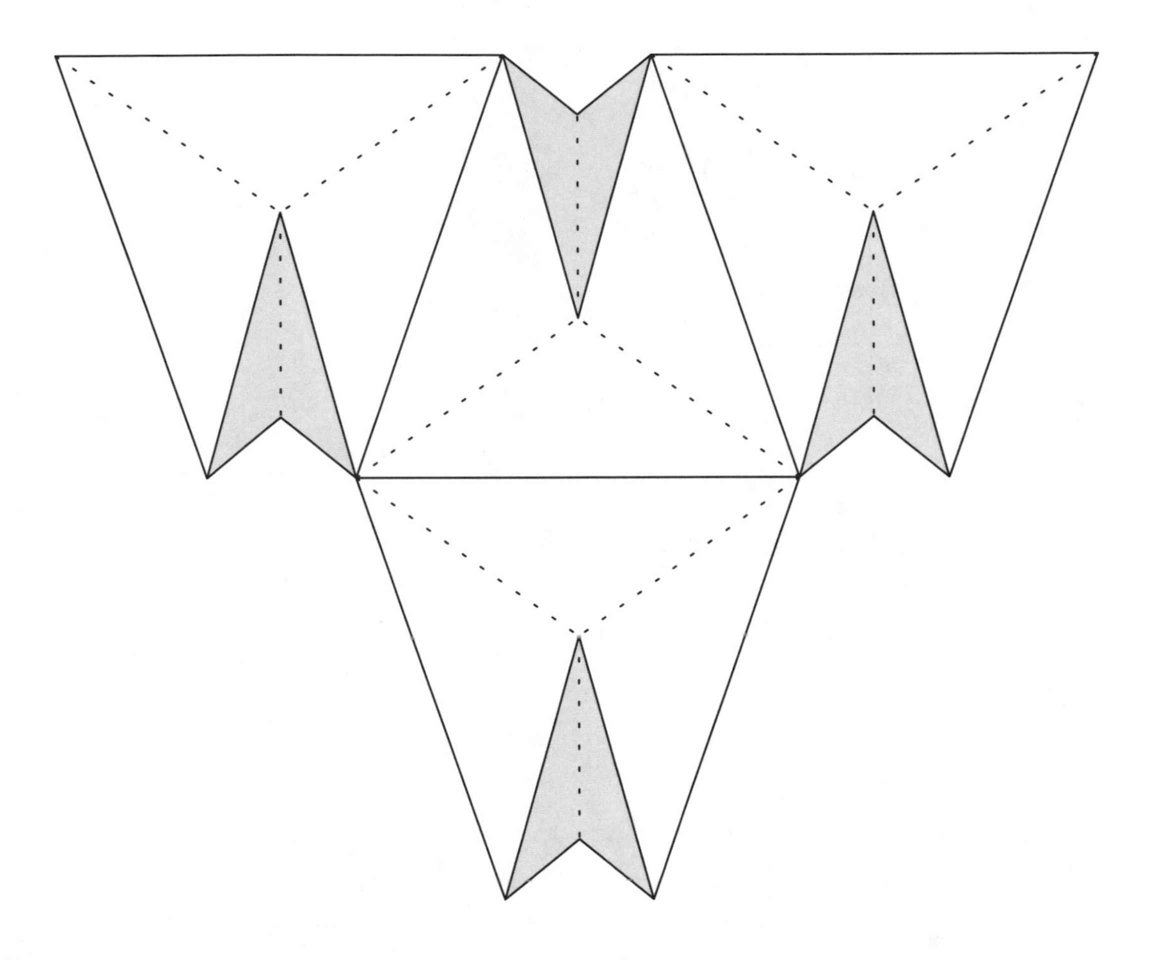

Tetrahedral molecules consist of four outer atoms distributed around a central atom. The basic tetrahedral model consists of four faces, which come together to form the bottom and three sides of a tetrahedron, as depicted in the introduction.

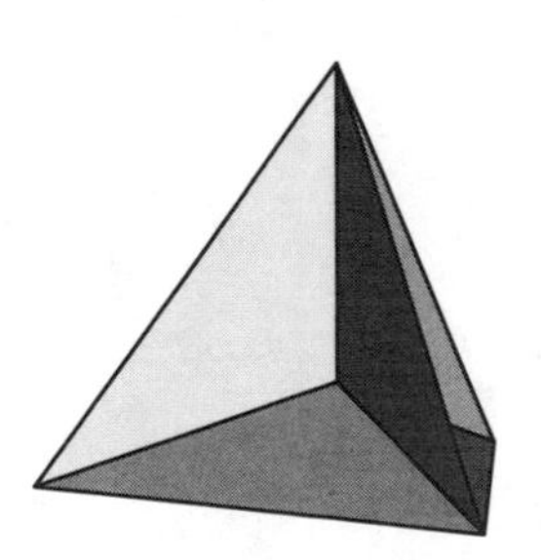

The angles X–A–X in the idealized model shown above and to the right are all 109.47°. In reality, these angles vary considerably, from 101.3° in POF_3 (page 65) to 119° in H_2SO_4 (page 71). Some of this variation is understandable using simple ideas. Some of it is the result of crystal packing forces, which tend to distort a molecule's "natural" tendencies.

In any case, it can be fun to try to rationalize the differences in related molecules, especially in relation to the AX_3E molecules presented in the previous section.

methane

shape: tetrahedral | units: pm | scale: 300,000,000:1

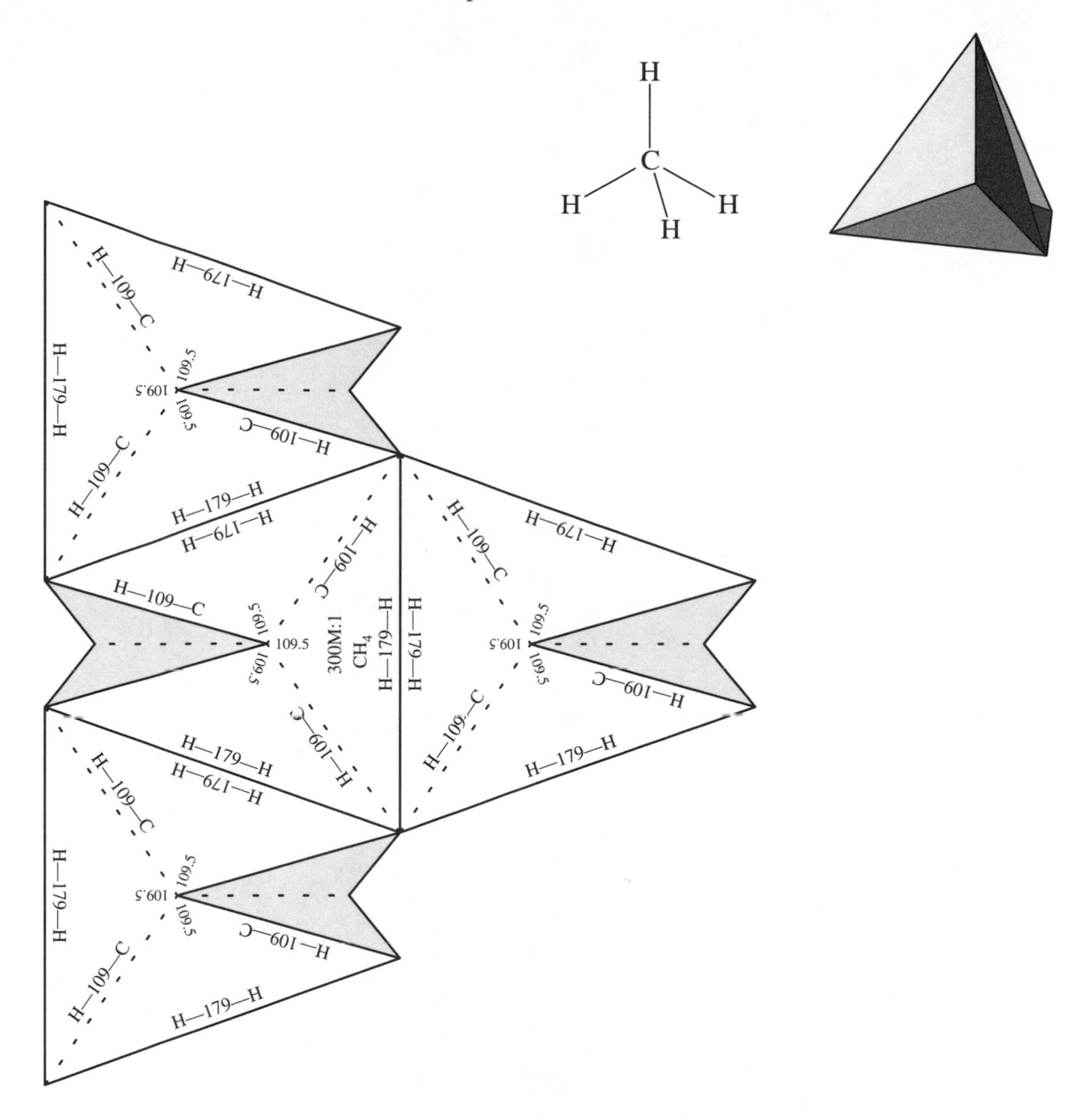

Questions to think about:

(a) Compare this structure to those of NH_4^+ (page 35) and BH_4^- (page 37), both of which also have 10 electrons. How do you explain the differences in size?

(b) In all of the simple derivatives of methane, the X–C–Y angle never varies by more than just a few degrees from 109.47°. What is so special about 109.47°?

(c) Methane is the classic tetrahedral molecule with "four equivalent bonds." But are they equivalent? Experiment indicates that they are not, even though all four of the C–H *connections* are identical. How can that be?

carbon tetrafluoride CF_4

shape: tetrahedral | units: pm | scale: 300,000,000:1

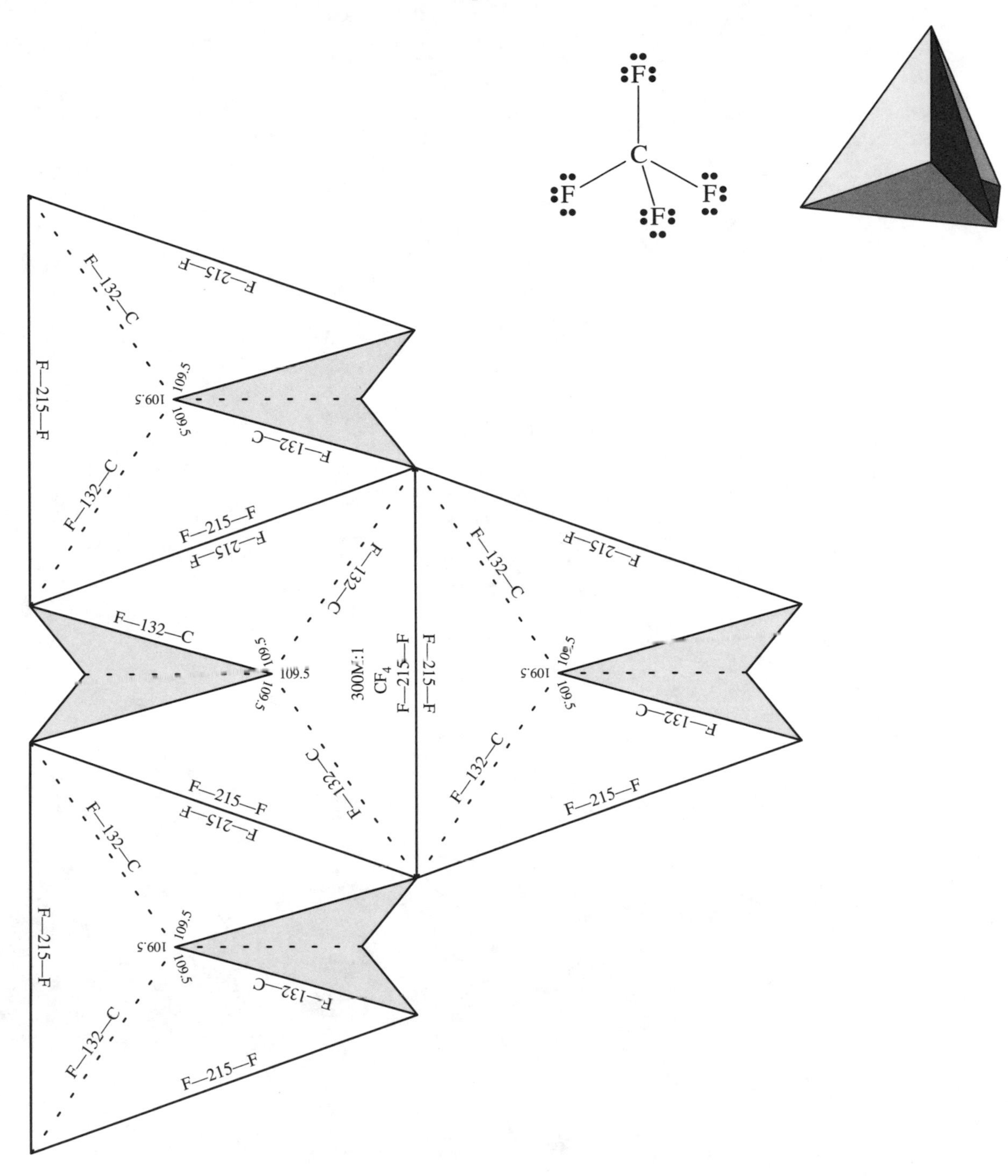

Questions to think about:

(a) How does this structure compare to that of CH_4 (page 31)?

(b) What would you predict for the structures of BF_4^- (page 39) and BeF_4^{-2}?

ammonium ion (in NH_4Br)

shape: tetrahedral units: pm scale: 300,000,000:1

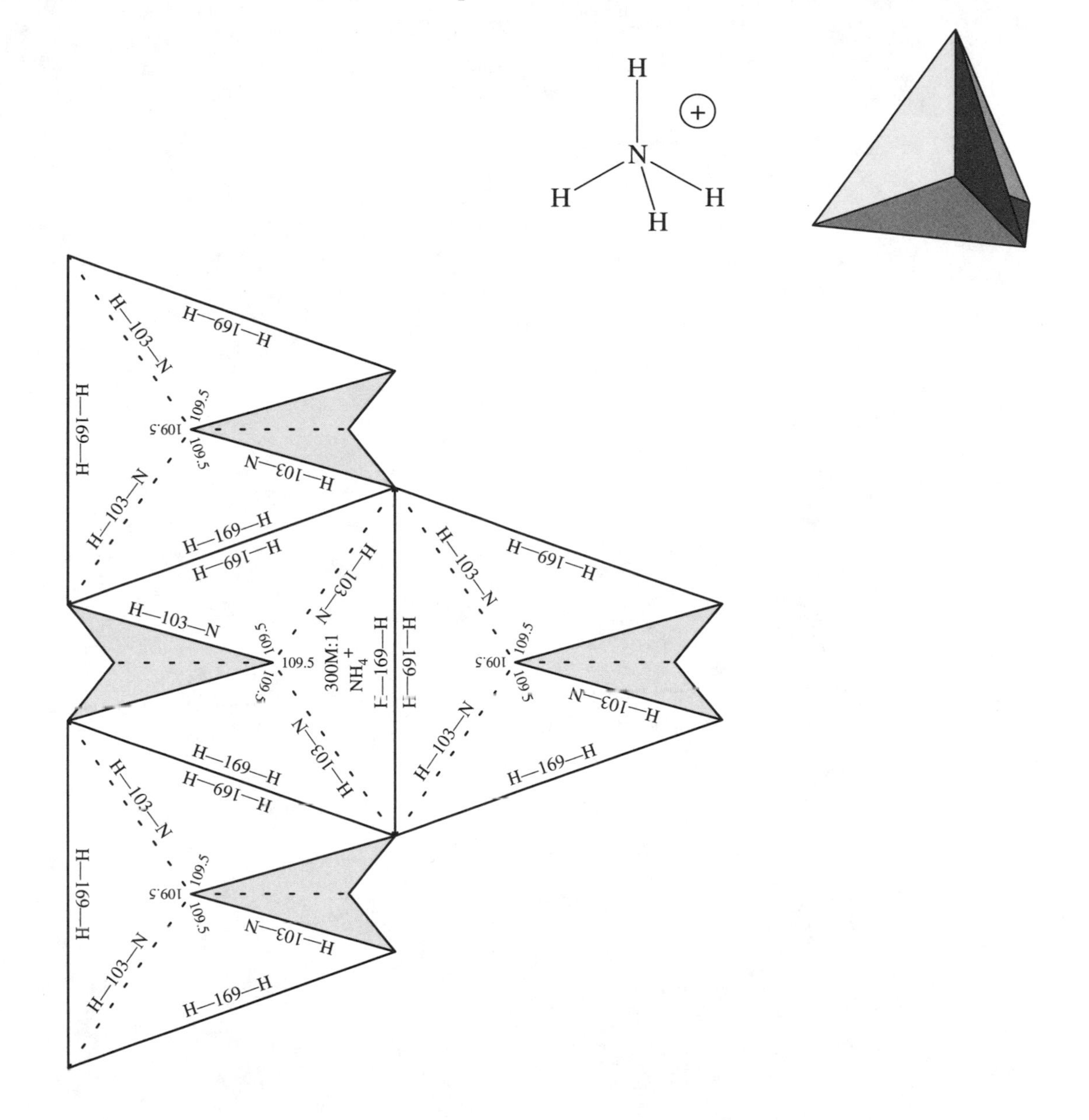

Question to think about:

The ammonium ion is extremely important in that it is a weak acid, undergoing the following equilibrium quite easily:

$$NH_4^+ + H_2O \rightleftharpoons H_3O^+ + NH_3$$

Which do you think is the *weaker* acid, H_3O^+ or NH_4^+? That is, which one do you think holds onto its protons more tightly? Why?

tetrahydroborate ion (in $NaBH_4$) BH_4^-

shape: tetrahedral units: pm scale: 300,000,000:1

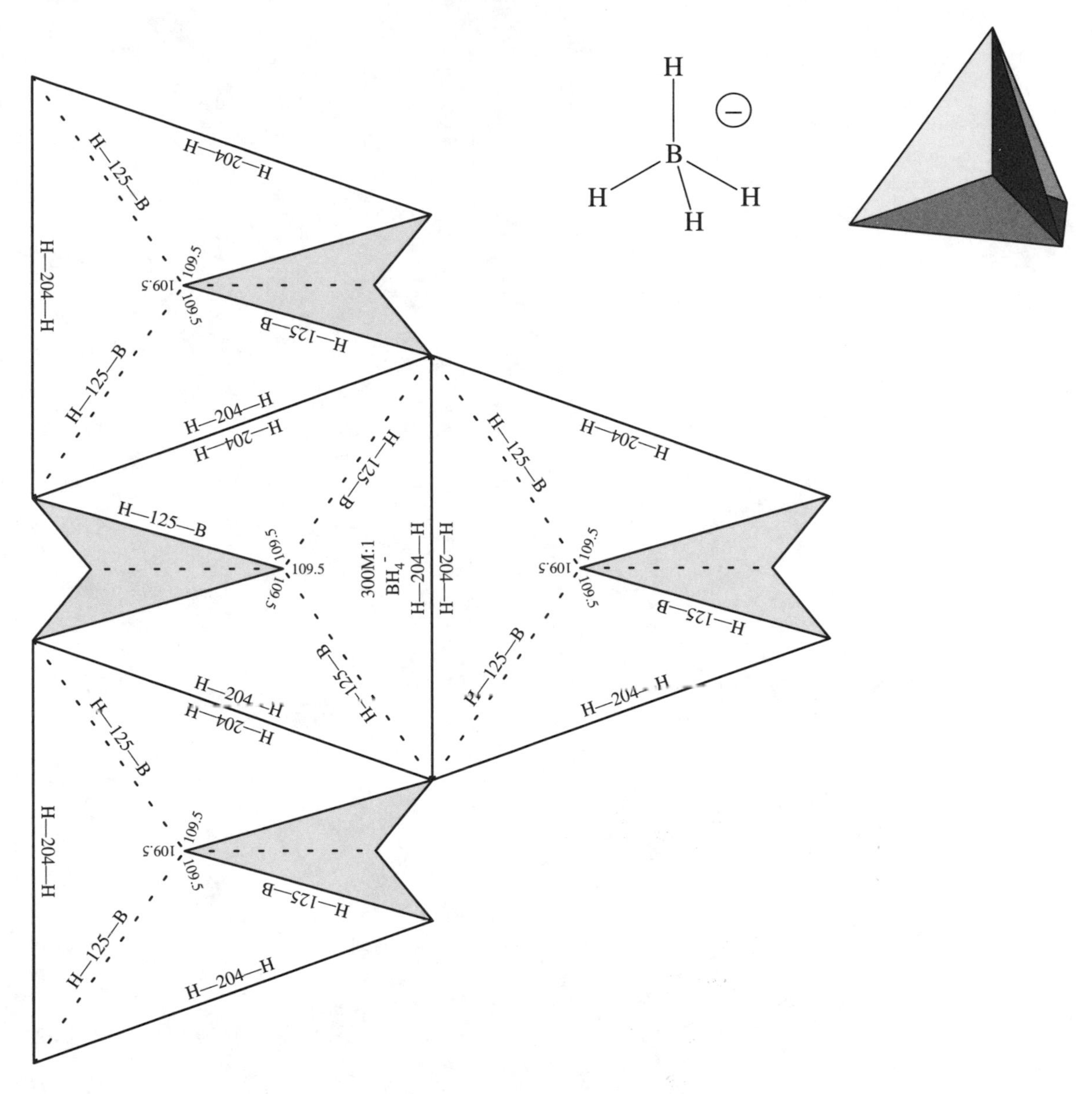

Questions to think about:

(a) How do you explain the observation that BH_4^- is so much larger than CH_4 (page 31)?

(b) "H^-" is called *hydride* and consists of two electrons and a proton. BH_4^- is considered a "hydride source" because in many situations it reacts by donating H^- to other molecules:

$$BH_4^- + X \longrightarrow \text{“}BH_3\text{”} + HX^-$$

"BH_3," like "H^-," doesn't actually exist. What is sometimes called BH_3 is really B_2H_6 (page 143). Why might H^- and BH_3 be so reactive as to not really exist?

tetrafluoroborate ion (in NH_4BF_4) BF_4^-

shape: tetrahedral units: pm scale: 300,000,000:1

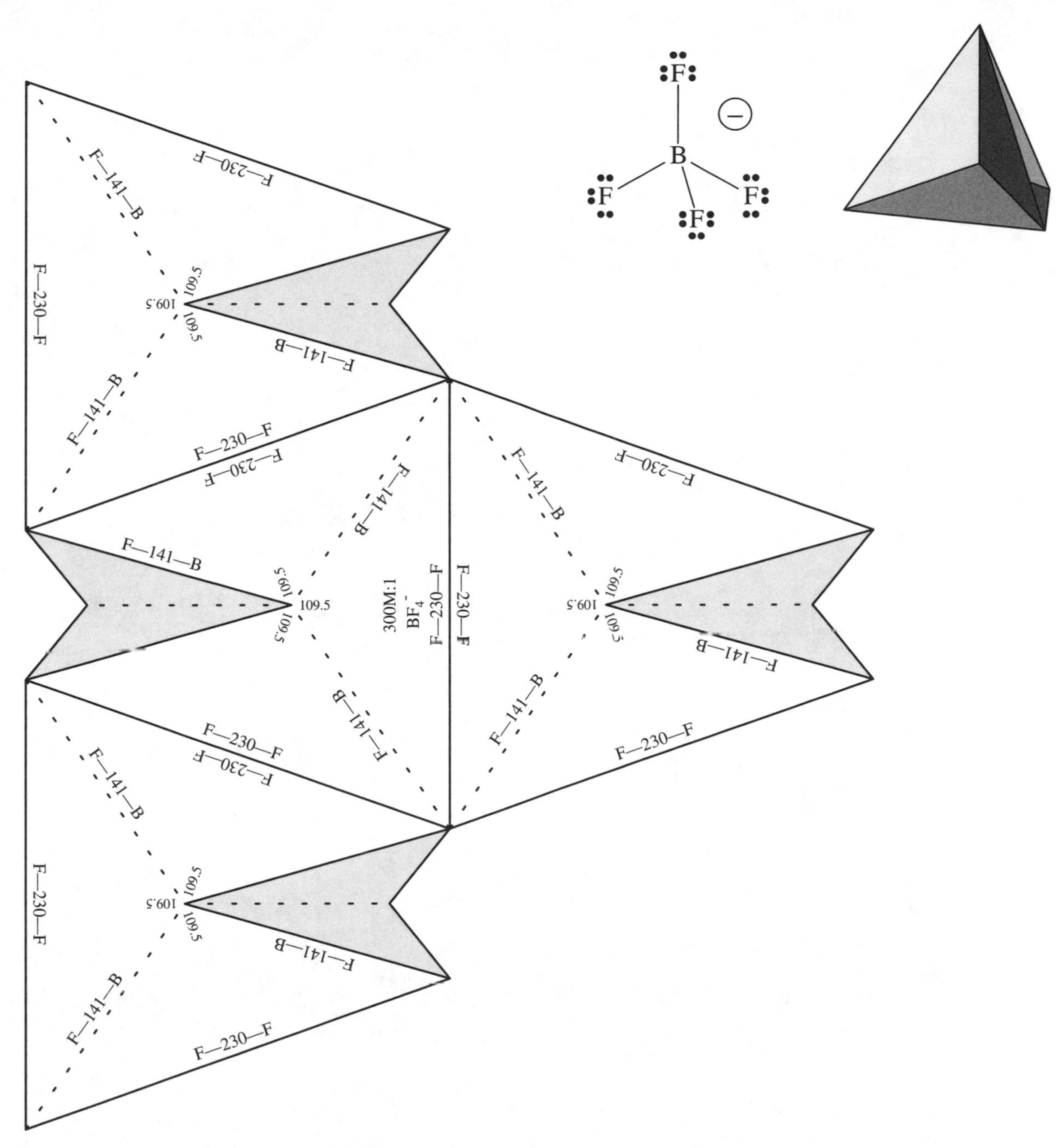

Questions to think about:

(a) Why is BF_4^- so much larger than BH_4^- (page 37)?

(b) Why is BF_4^- so much smaller than BeF_4^{-2} (Be–F 157 pm)?

(c) Compare the structure of BF_4^- to that of BF_3 (see page 13). How do you explain the differences?

boron trifluoride-ammonia complex $BF_3{\cdot}NH_3$

shape: tetrahedral units: pm scale: 300,000,000:1

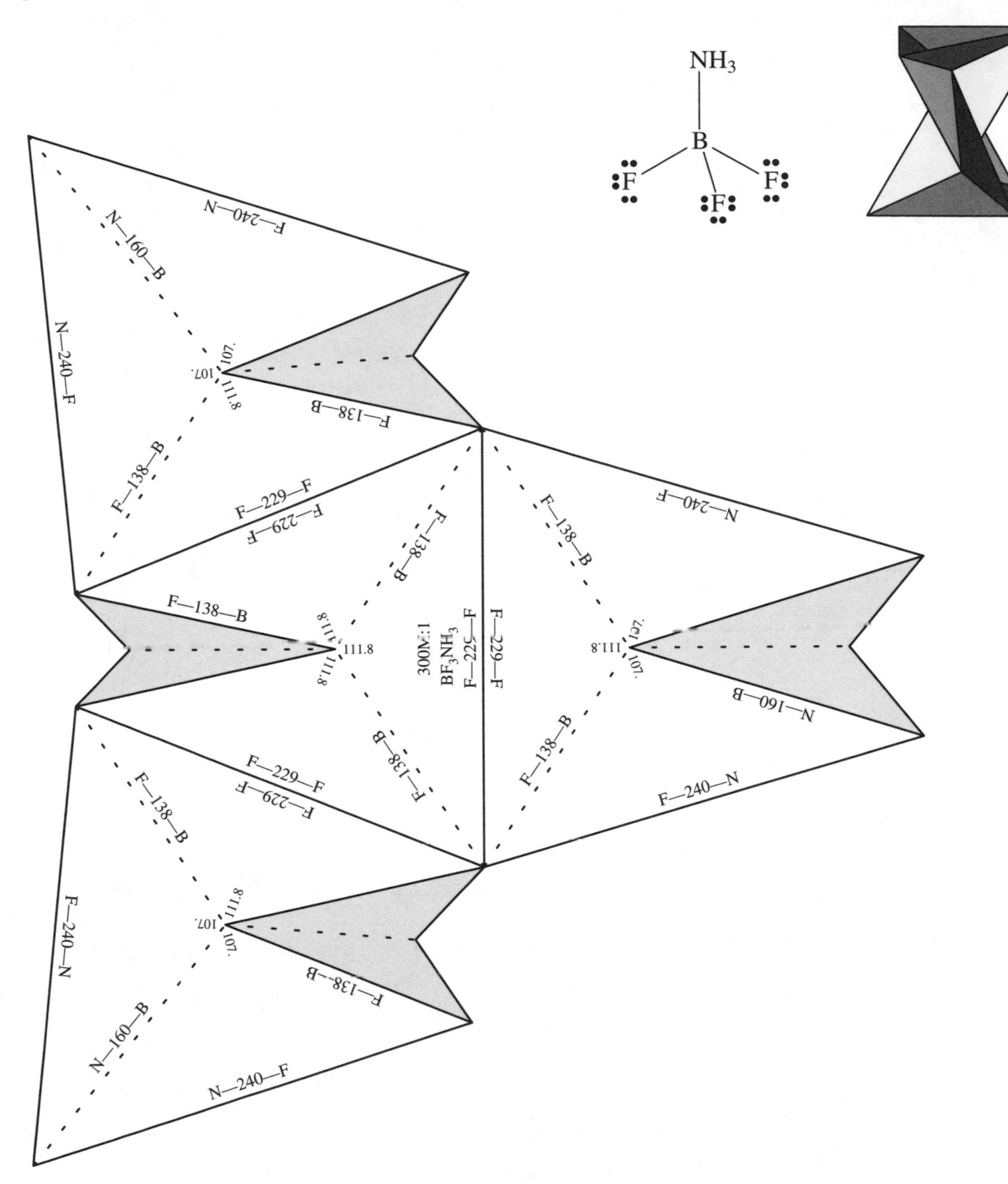

Questions to think about:

(a) *Lewis bases* donate pairs of electrons when reacting; *Lewis acids* accept them. This complex is the classic example of the result of a Lewis acid/base interaction. What was the Lewis acid and what was the Lewis base?

(b) The N–H distances is $BF_3{\cdot}NH_3$ have not been precisely determined experimentally. Would you expect them to be larger or smaller than the N–H distances in NH_3?

(c) In comparison to BF_3 (see page 13), the B–F distances in $BF_3{\cdot}NH_3$ are significantly longer. Why?

borane-phosphorus trifluoride complex (B) $BH_3 \cdot PF_3$

shape: tetrahedral units: pm scale: 300,000,000:1

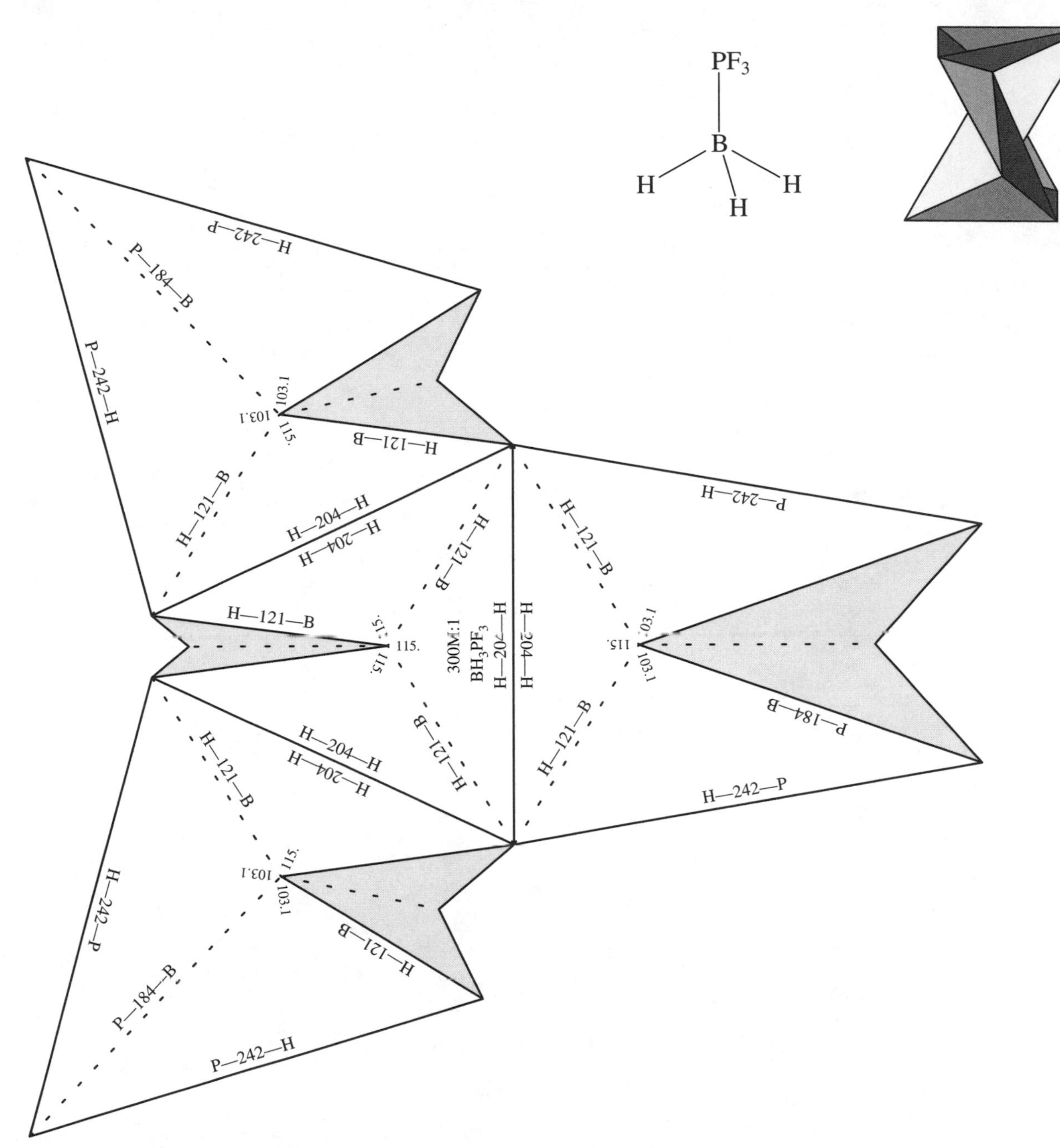

Note: The other half of $BH_3 \cdot PF_3$ is given on page 45. These two models may be spliced together to form the whole molecule.

Questions to think about:

(a) Although BH_3 does not exist by itself, here is an example of one of its stable complexes. Can you write an equation making this complex from B_2H_6 and PF_3?

(b) In the analogous structure, CH_3–SiF_3, the C–Si distance is known to be 188 pm. How do you explain the smaller distance of 184 pm for the B–P distance in $BH_3 \cdot PH_3$?

shape: tetrahedral units: pm scale: 300,000,000:1

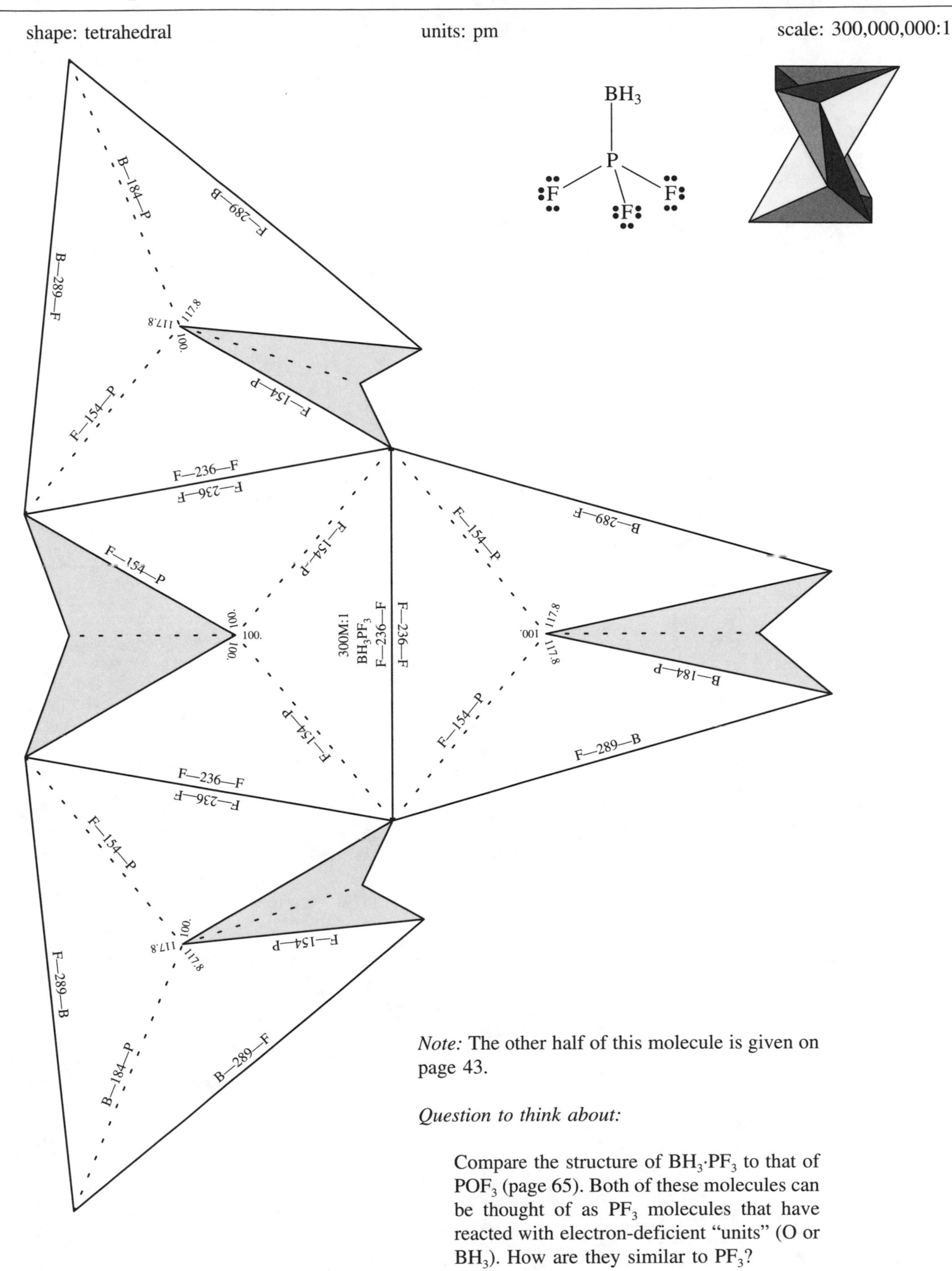

Note: The other half of this molecule is given on page 43.

Question to think about:

Compare the structure of $BH_3 \cdot PF_3$ to that of POF_3 (page 65). Both of these molecules can be thought of as PF_3 molecules that have reacted with electron-deficient "units" (O or BH_3). How are they similar to PF_3?

borane-carbon monoxide complex $BH_3{\cdot}CO$

shape: tetrahedral units: pm scale: 300,000,000:1

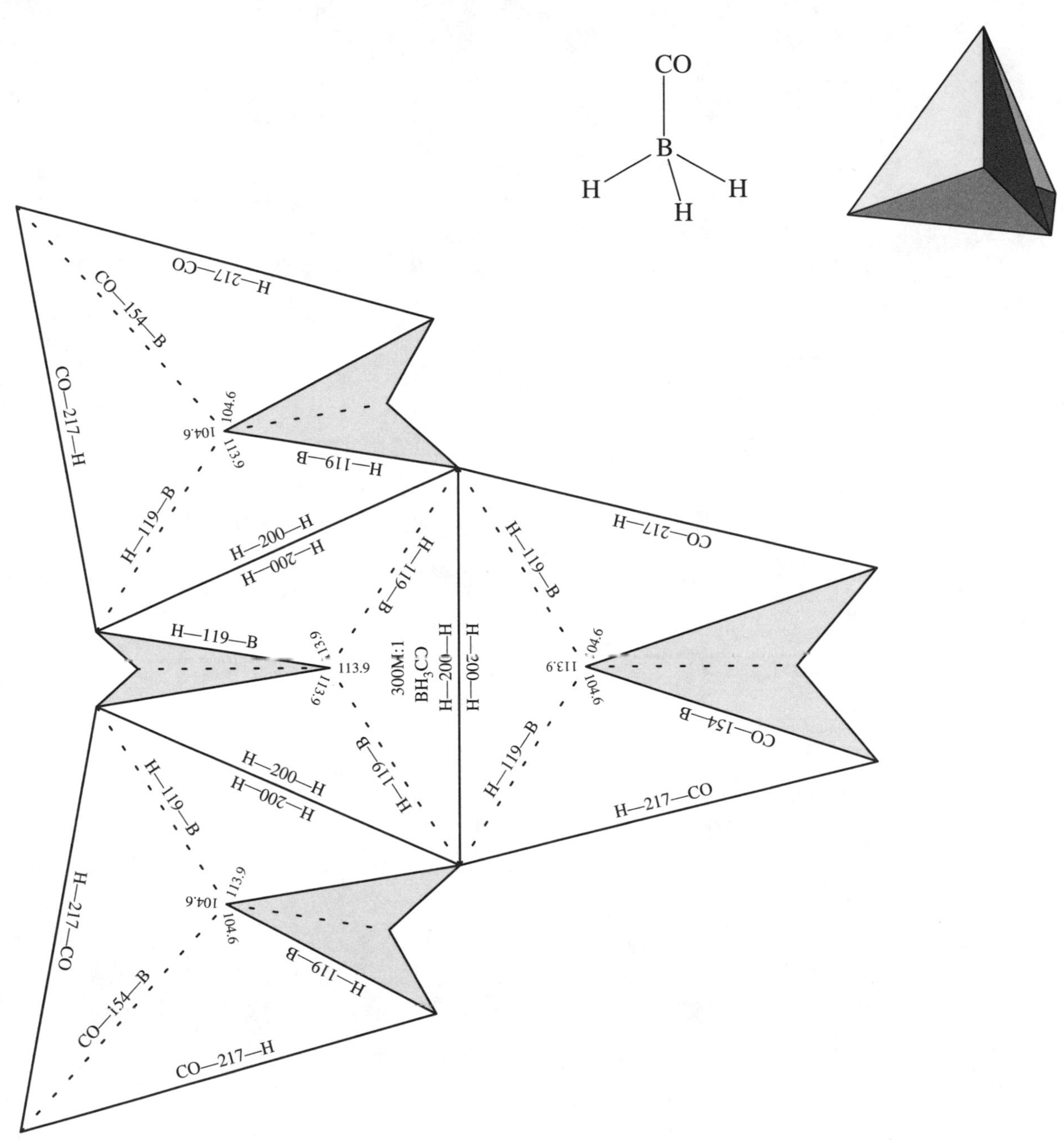

Questions to think about:

(a) Would you expect the B–H distances in the hypothetical BH_3 molecule to be larger or smaller than the B–H distances in $BH_3{\cdot}CO$?

(b) The C–O distance in $BH_3{\cdot}CO$ is 113 pm. This is exactly the distance in carbon monoxide itself. Why might this be?

shape: tetrahedral units: pm scale: 300,000,000:1

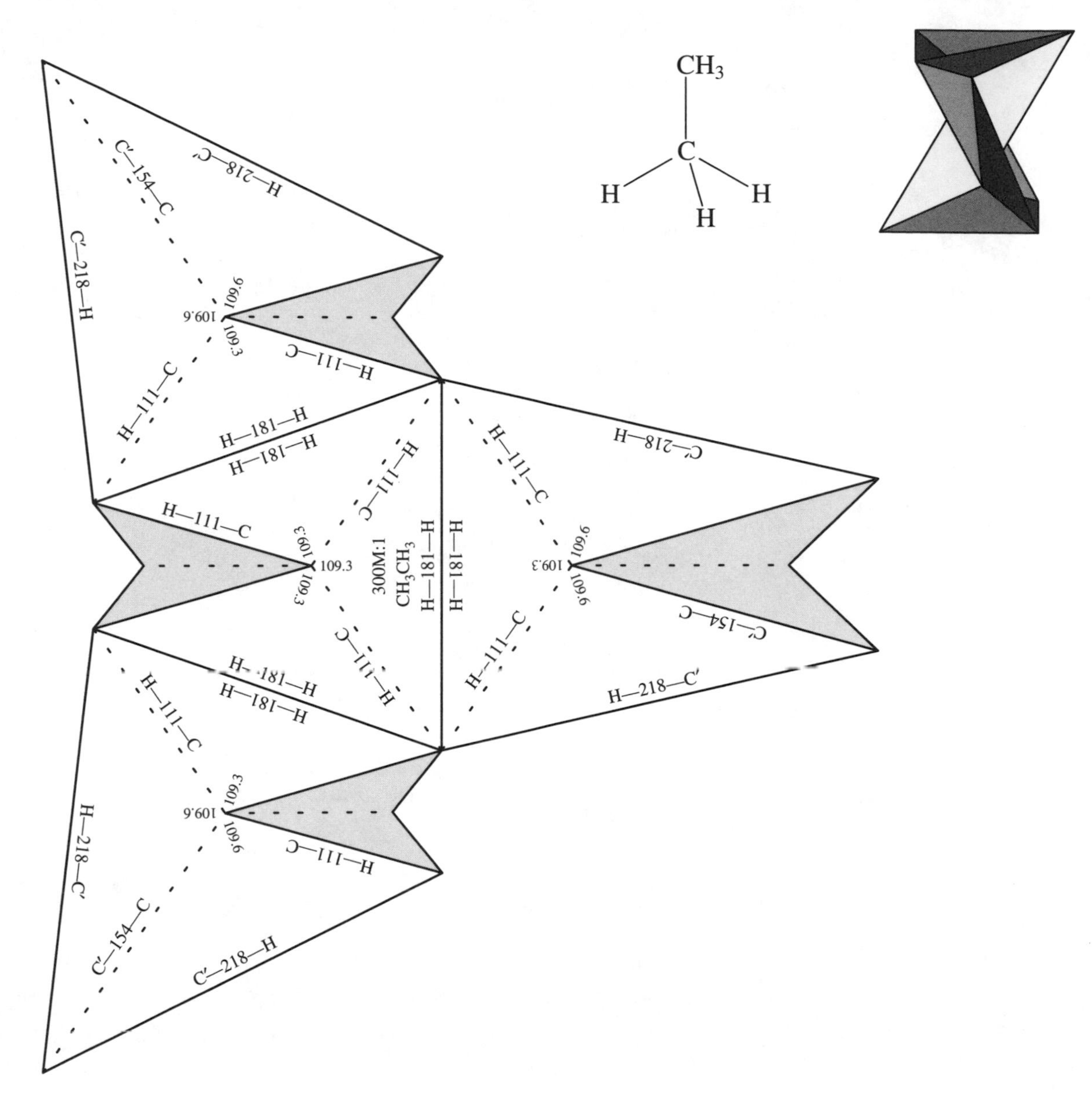

Note: This is half of CH_3CH_3. Make two and splice them together to form the whole molecule. Looking down the C–C axis you should see the view shown to the right, called the *staggered conformation*. This view down the C–C axis is called the *Newman projection.*

Question to think about:

The alternative conformation for ethane is called *eclipsed*, in which the front CH_3 in the Newman projection is turned 60°. In the eclipsed conformation the three front H atoms line up with the three H atoms in the back. Why might the staggered conformation be more stable than the eclipsed conformation?

shape: tetrahedral units: pm scale: 300,000,000:1

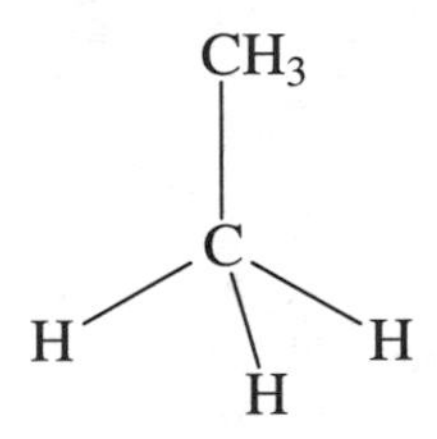

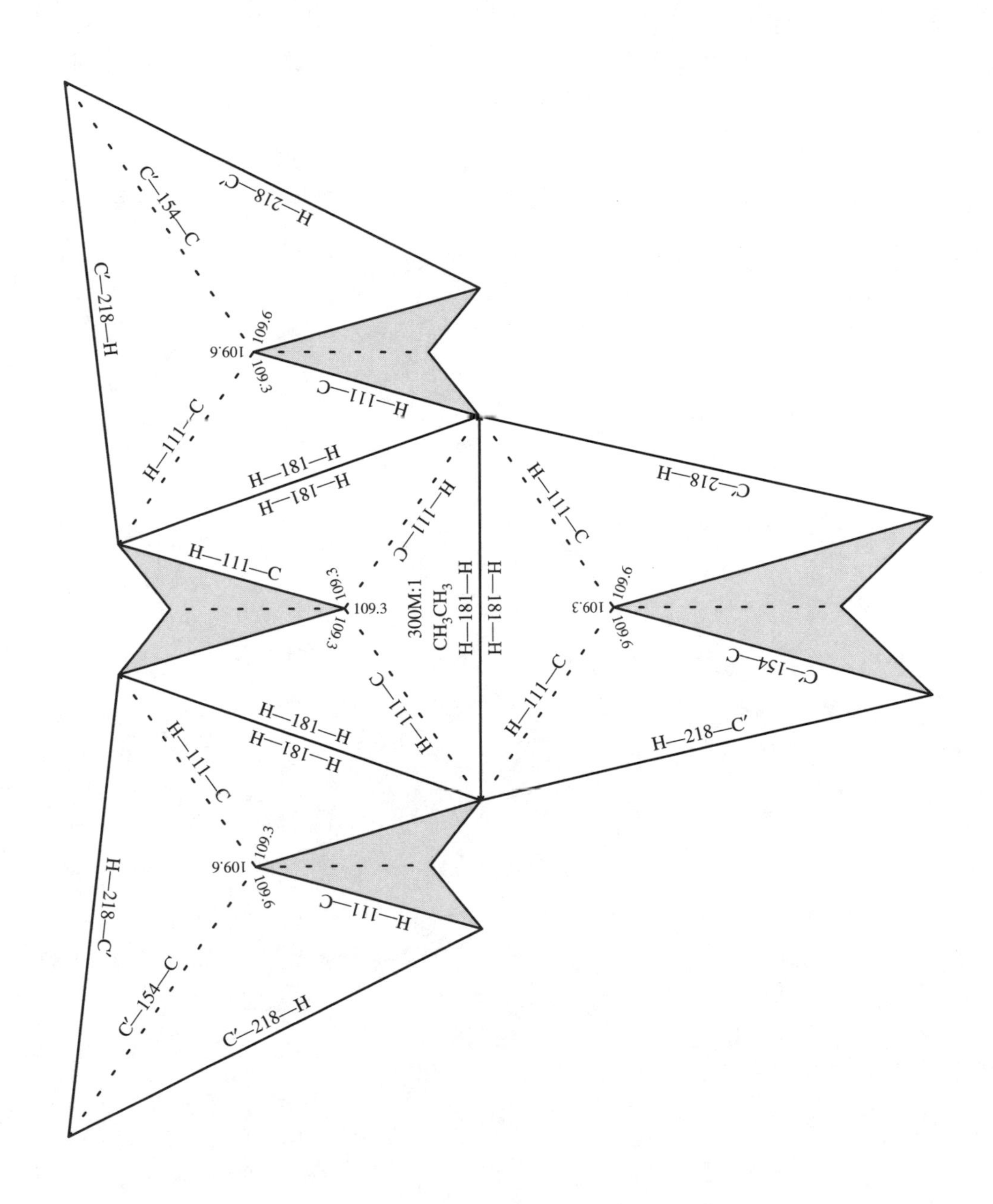

methylamine (C) CH_3NH_2

shape: tetrahedral units: pm scale: 300,000,000:1

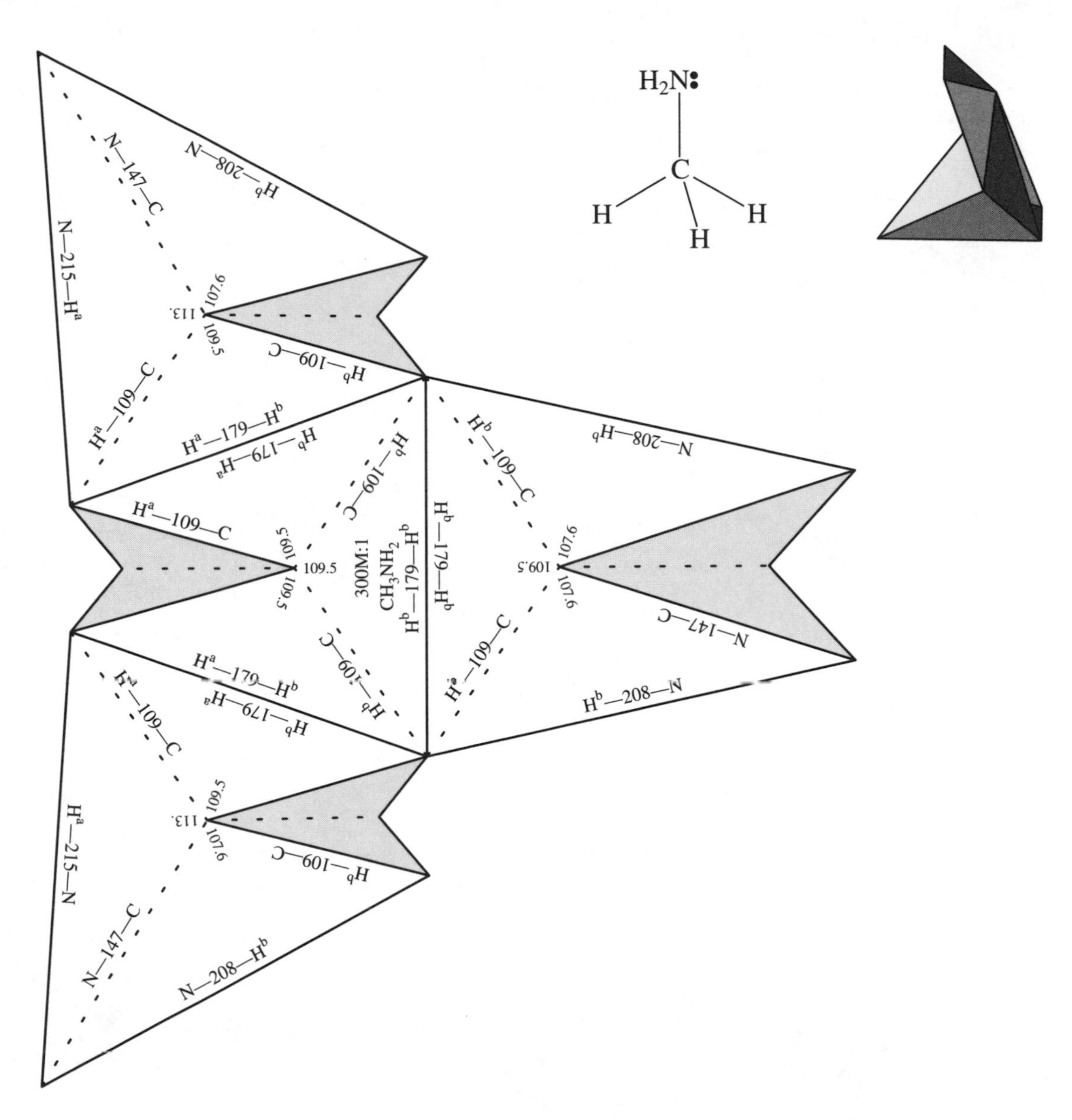

Note: The other half of CH_3NH_2 is given on page 17. These two models may be spliced together to form the whole molecule. The C–Hᵃ line should bisect the H–N–H angle. The conformation is thus *staggered*, as for CH_3CH_3 (page 49).

Questions to think about:

(a) If you look carefully at this model you will notice that the H–C–N angles are not all the same. The one to Hᵃ is 113°, while the ones to Hᵇ are only 107.9°. That is, the NH_2 group leans away from Hᵃ. What is going on here?

(b) Which is flatter around the nitrogen atom, CH_3NH_2 (see page 17) or NH_3 (page 11)? What does this mean for the lone pair?

(c) Which would you predict to be more basic, CH_3NH_2 or NH_3?

methanol CH_3OH

shape: tetrahedral units: pm scale: 300,000,000:1

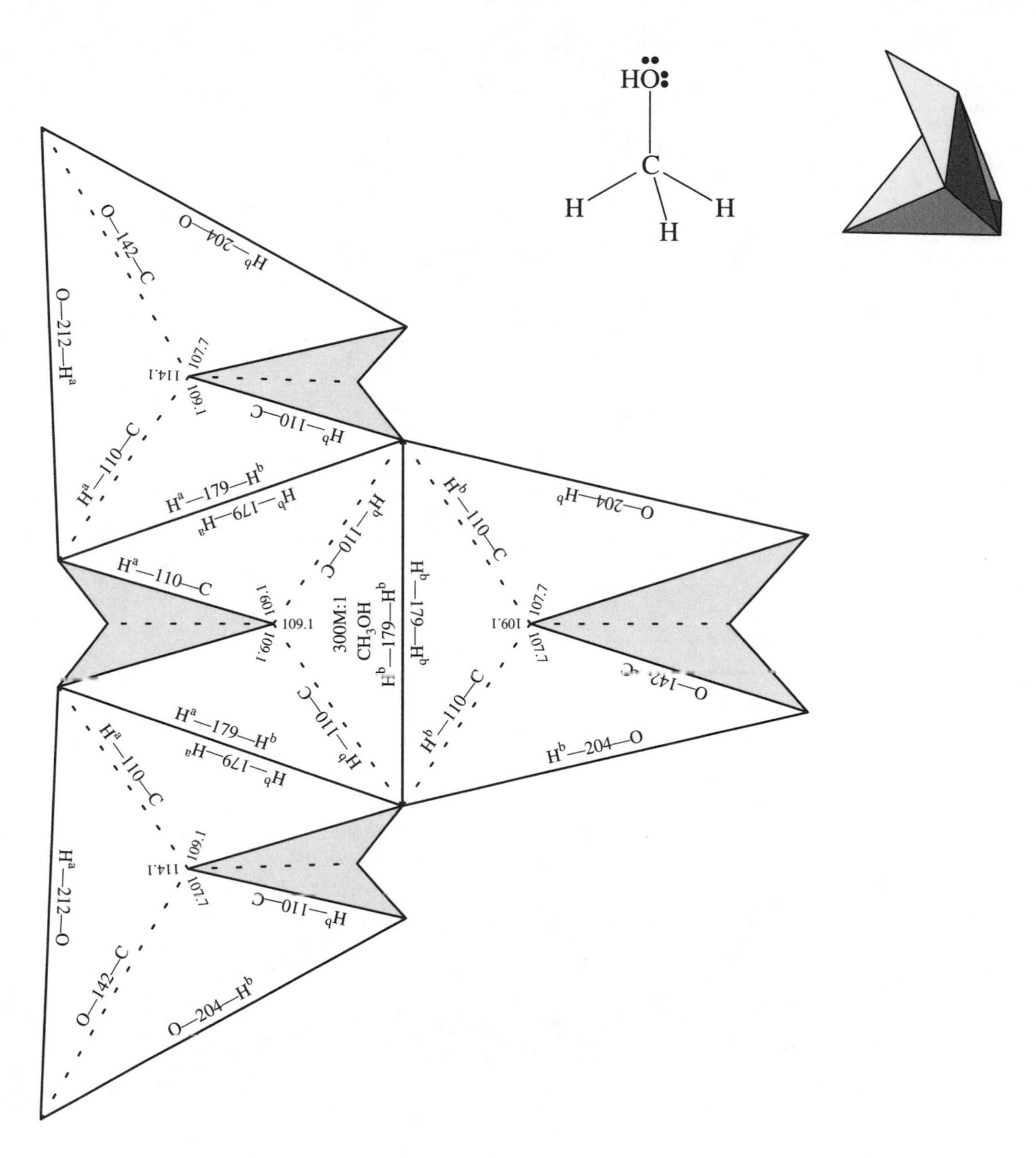

Note: The other half of CH_3OH is given to the right (doubled so that writing appears on both sides once folded). Fold these two models together. The C–H_a line should *eclipse* the O–H line.

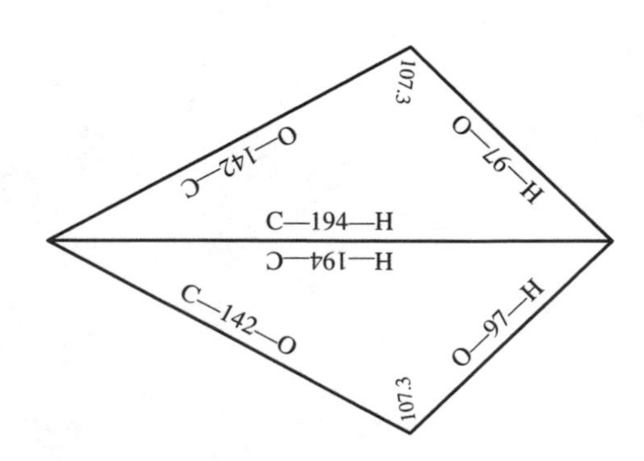

Questions to think about:

(a) Once again, careful examination of the model indicates that the OH leans away from one of the hydrogen atoms. Why might this be the case?

(b) Why is the C–O distance in CH_3OH smaller than the C–N distance in CH_3NH_2 (page 53)?

fluoroform CHF_3

shape: tetrahedral units: pm scale: 300,000,000:1

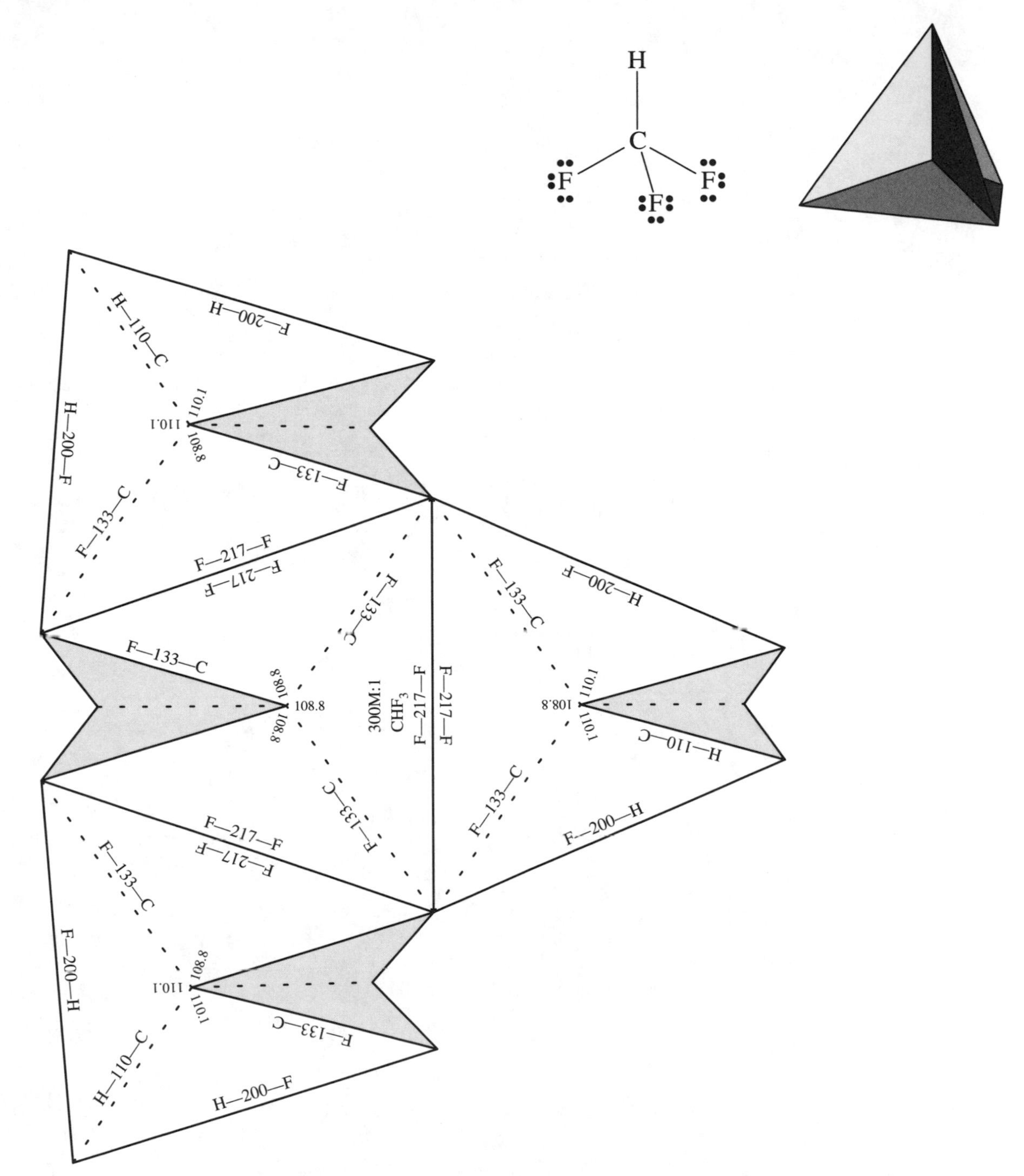

Question to think about:

How much difference would you expect there to be between this structure and that of $SiHF_3$?

chloroform

shape: tetrahedral | units: pm | scale: 300,000,000:1

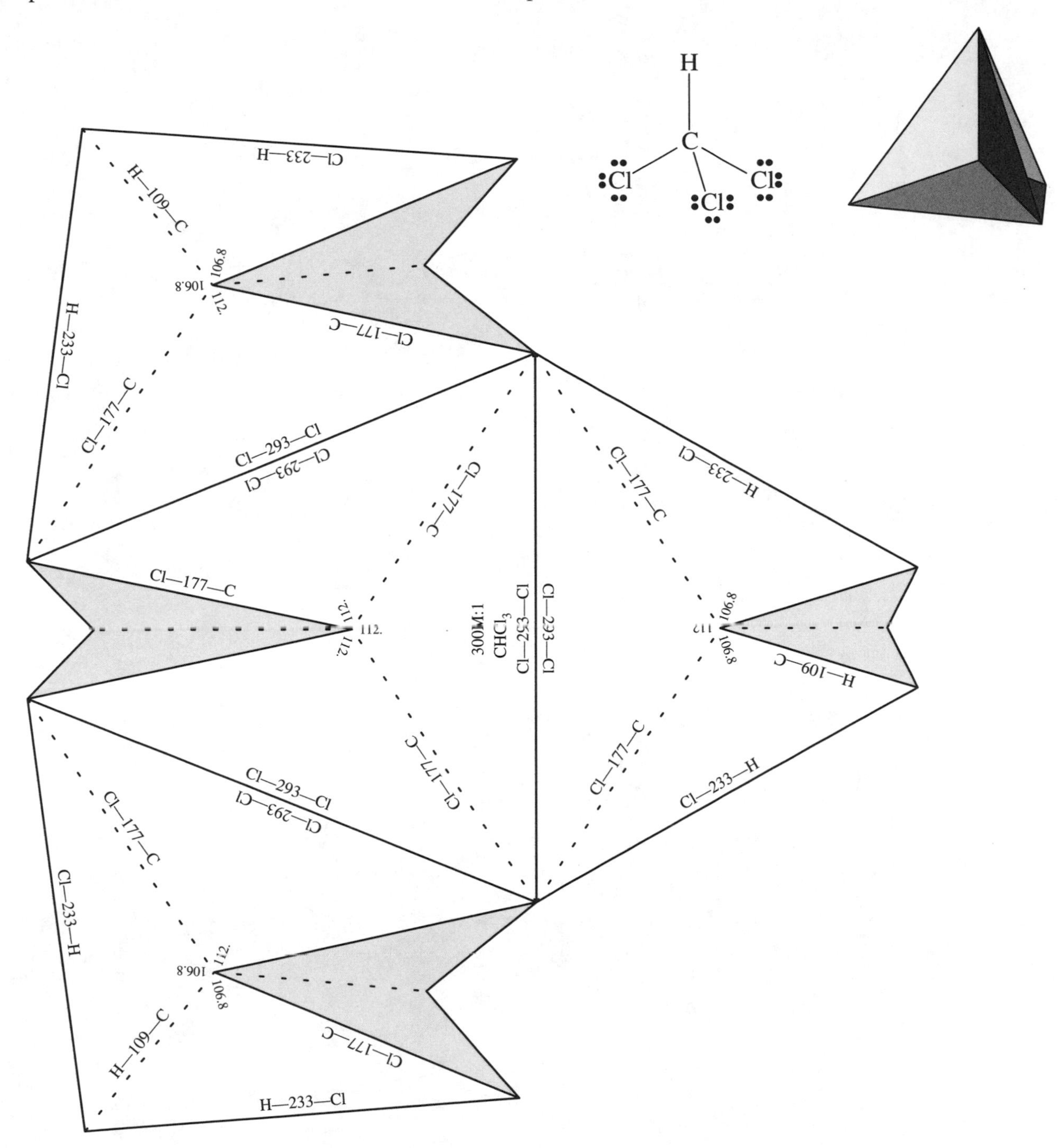

Questions to think about:

(a) Chloroform is weakly acidic ($pK_a \approx 25$) and reacts with strong bases such as NaOH to form CCl_3^-. How are NCl_3 and CCl_3^- similar?

(b) What is the probable structure (shape, distances, angles) of CCl_3^-?

silane SiH_4

shape: tetrahedral units: pm scale: 300,000,000:1

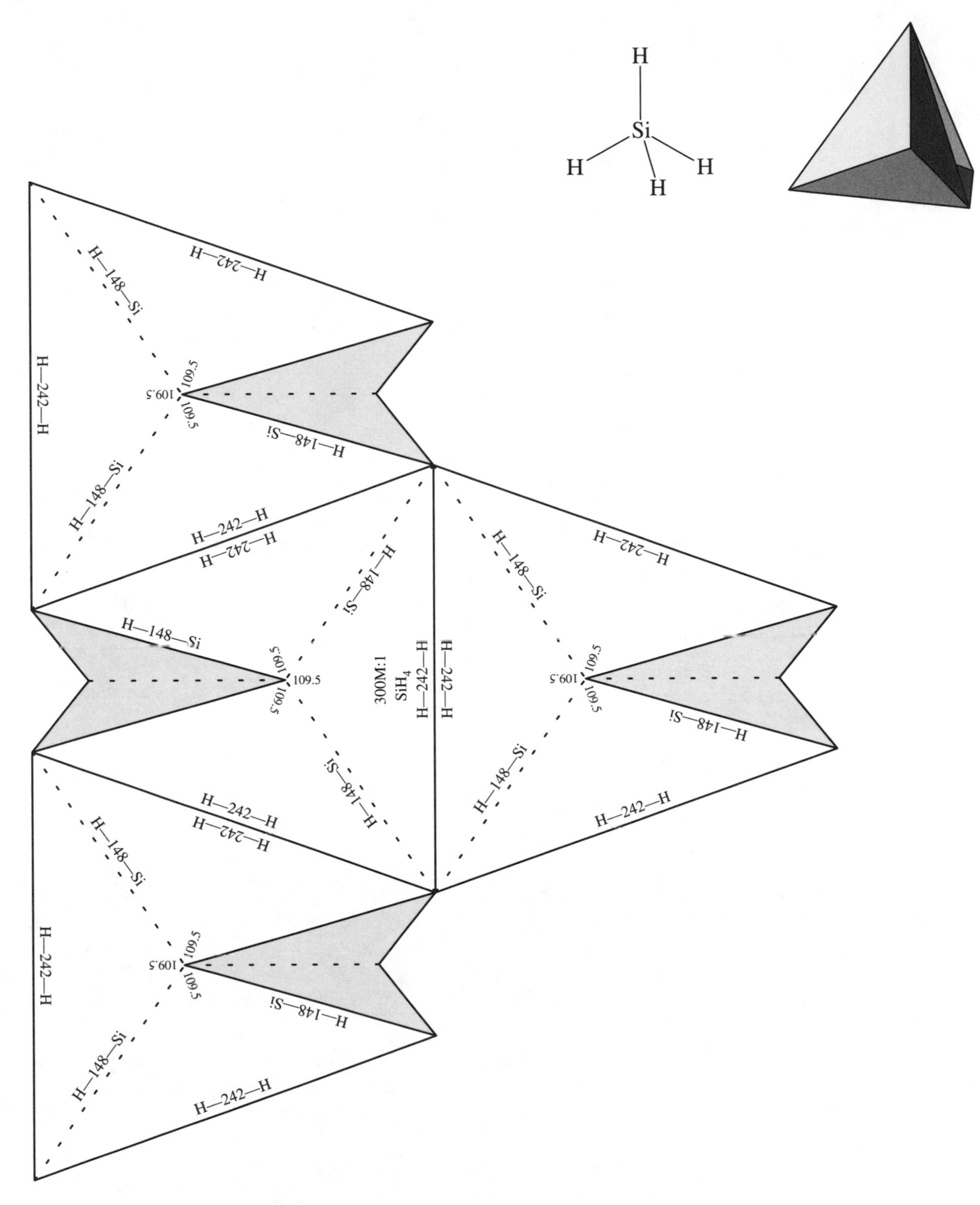

Note how much larger SiH_4 is than CH_4 (page 31)!

silicon tetrafluoride  SiF_4

shape: tetrahedral   units: pm   scale: 300,000,000:1

Question to think about:

Imagine that one of the protons in one of the F atoms were somehow moved over to join with the Si atom. What molecule would you now have? What do you predict for its structure? (See page 65.)

phosphorus oxyfluoride POF_3

shape: tetrahedral units: pm scale: 300,000,000:1

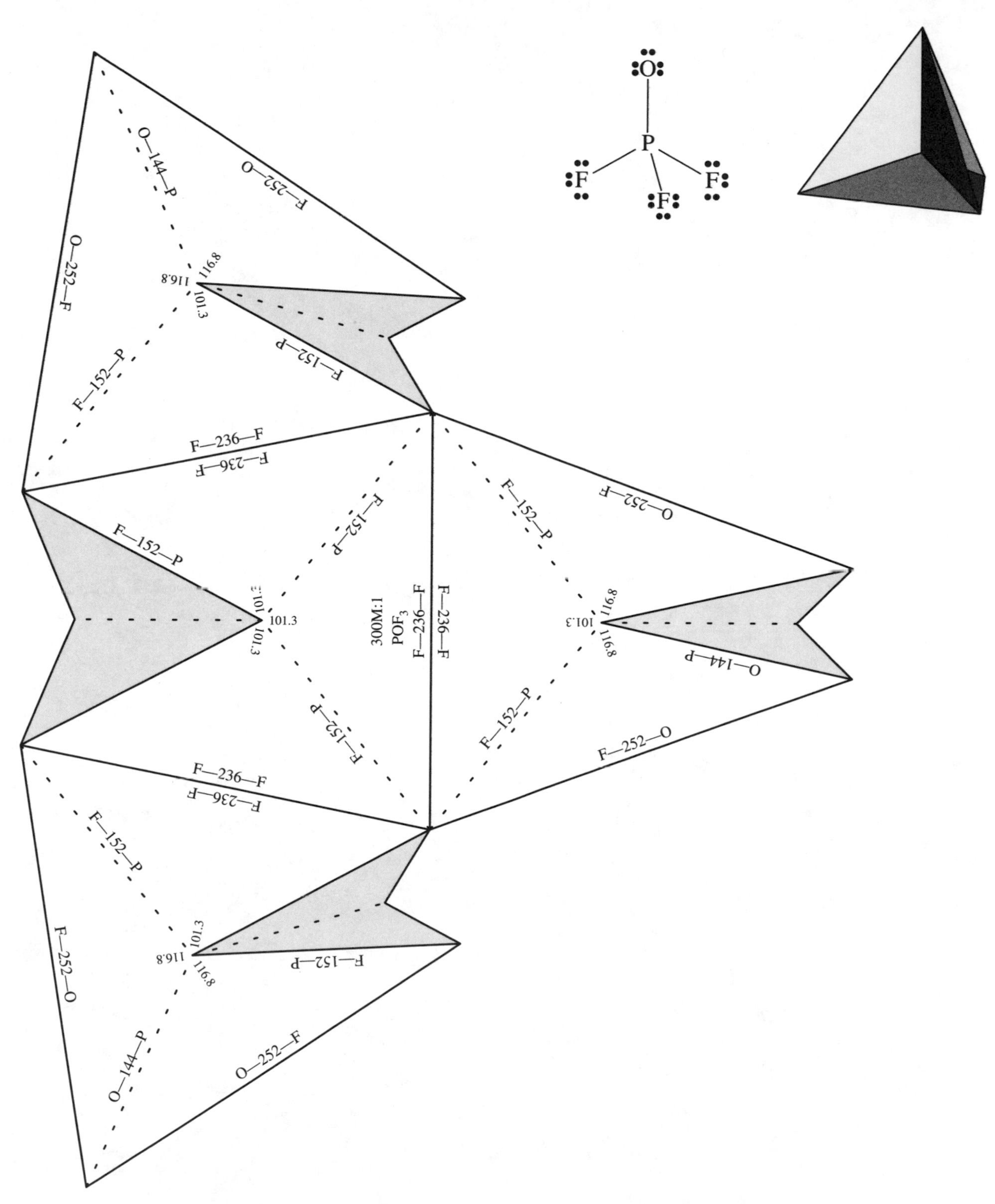

Question to think about:

It is interesting to compare the structures of PF_3 and POF_3. What other molecules and ions in this book are similarly related by oxidation? Do they show the same trends?

phosphoric acid (a) H_3PO_4

shape: tetrahedral | units: pm | scale: 300,000,000:1

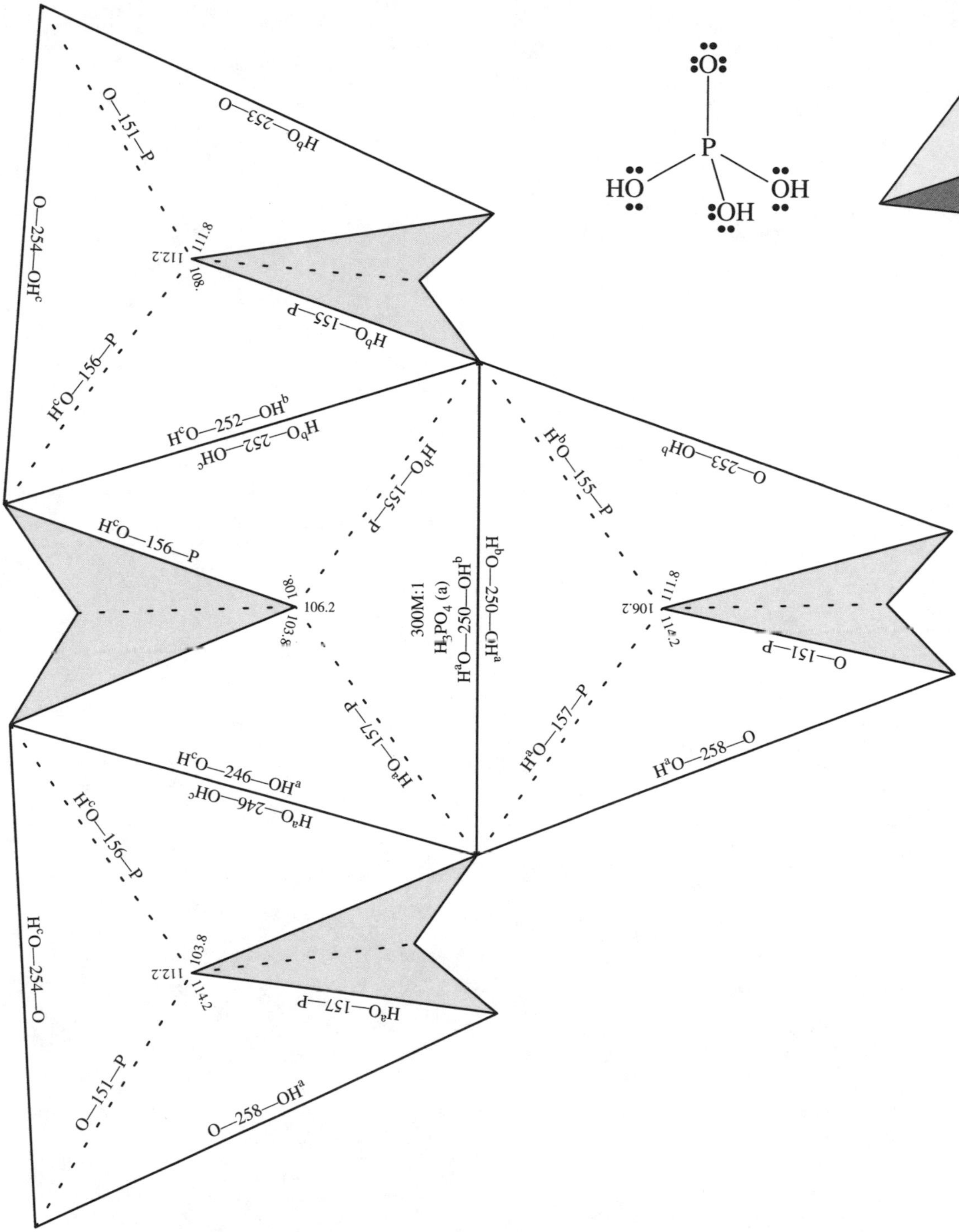

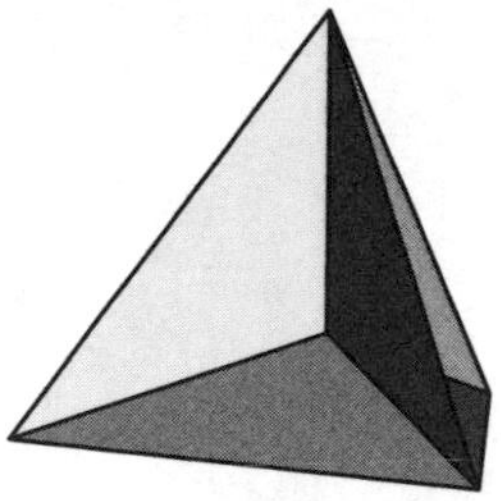

Questions to think about:

(a) This structure was based on x-ray analysis of a crystal. If you look carefully, you will see that there are three *different* OH groups, each with slightly different angles to the P–O bond. What could cause this variation?

(b) Compare this structure to that of H_2SO_4 (page 71). Why is the S–O distance smaller than the P–O distance?

phosphoric acid (b) H_3PO_4

shape: tetrahedral | units: pm | scale: 300,000,000:1

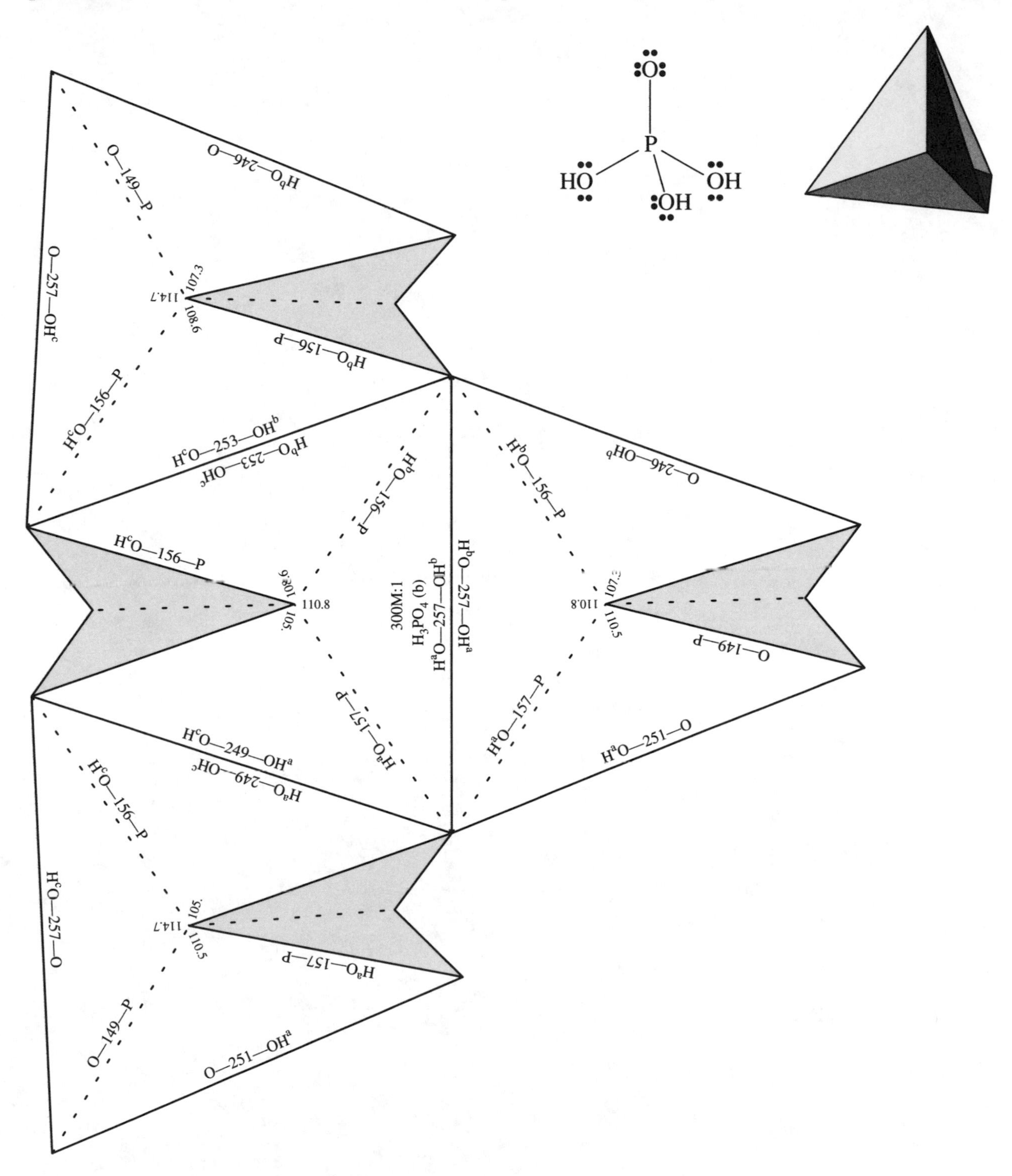

Questions to think about:

(a) This is a slightly different structure from the previous one, also found in the same crystal. What lesson can we learn from this comparison?

(b) Why would the P–O distance in H_3PO_4 be larger than the P–O distance in POF_3 (page 65)?

sulfuric acid H_2SO_4

shape: tetrahedral units: pm scale: 300,000,000:1

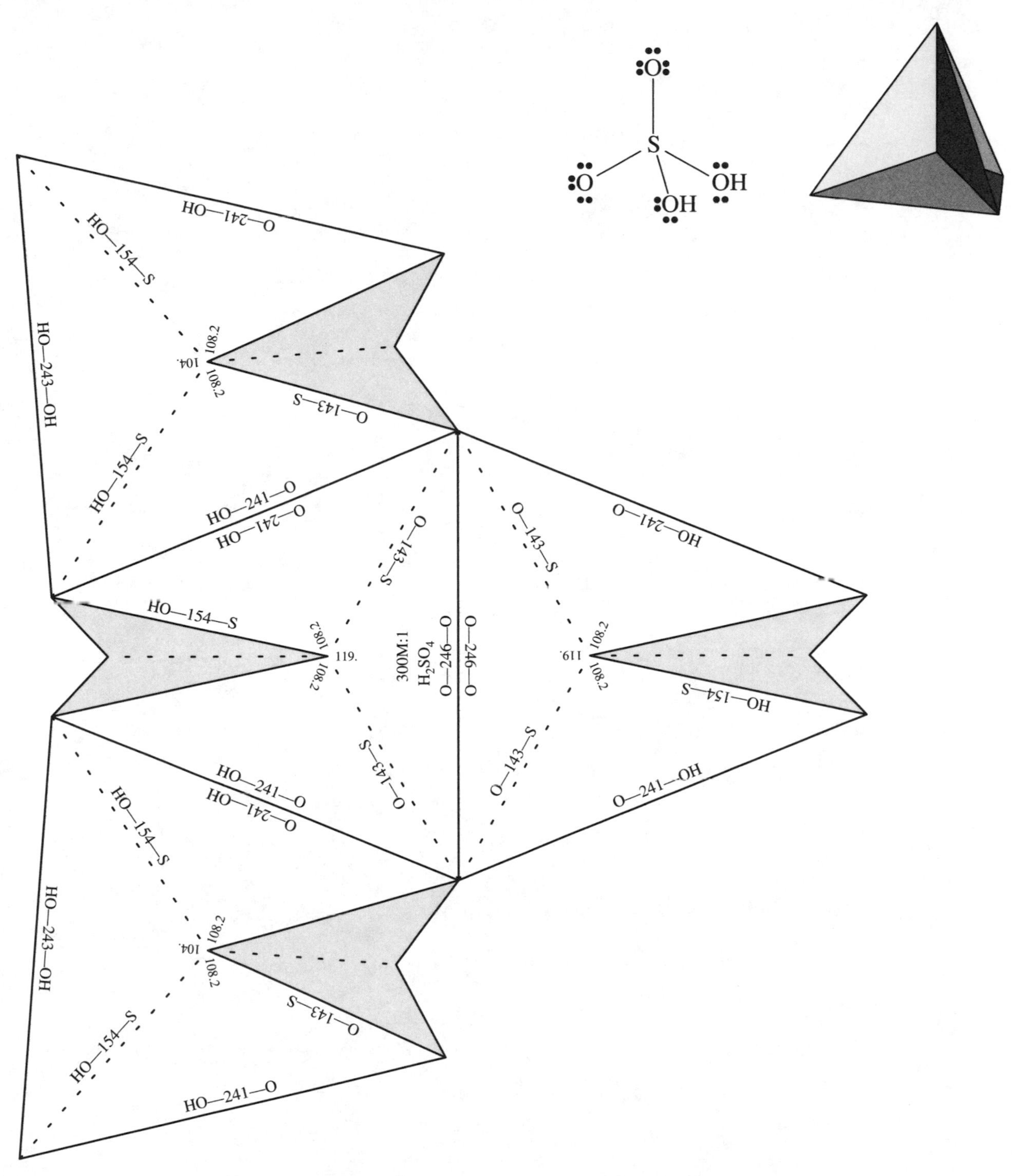

Question to think about:

Why might the O–S–O angle for the non-OH oxygens in H_2SO_4 be so large compared to any of the other angles?

perchlorate ion (in NH_4ClO_4) ClO_4^-

shape: tetrahedral units: pm scale: 300,000,000:1

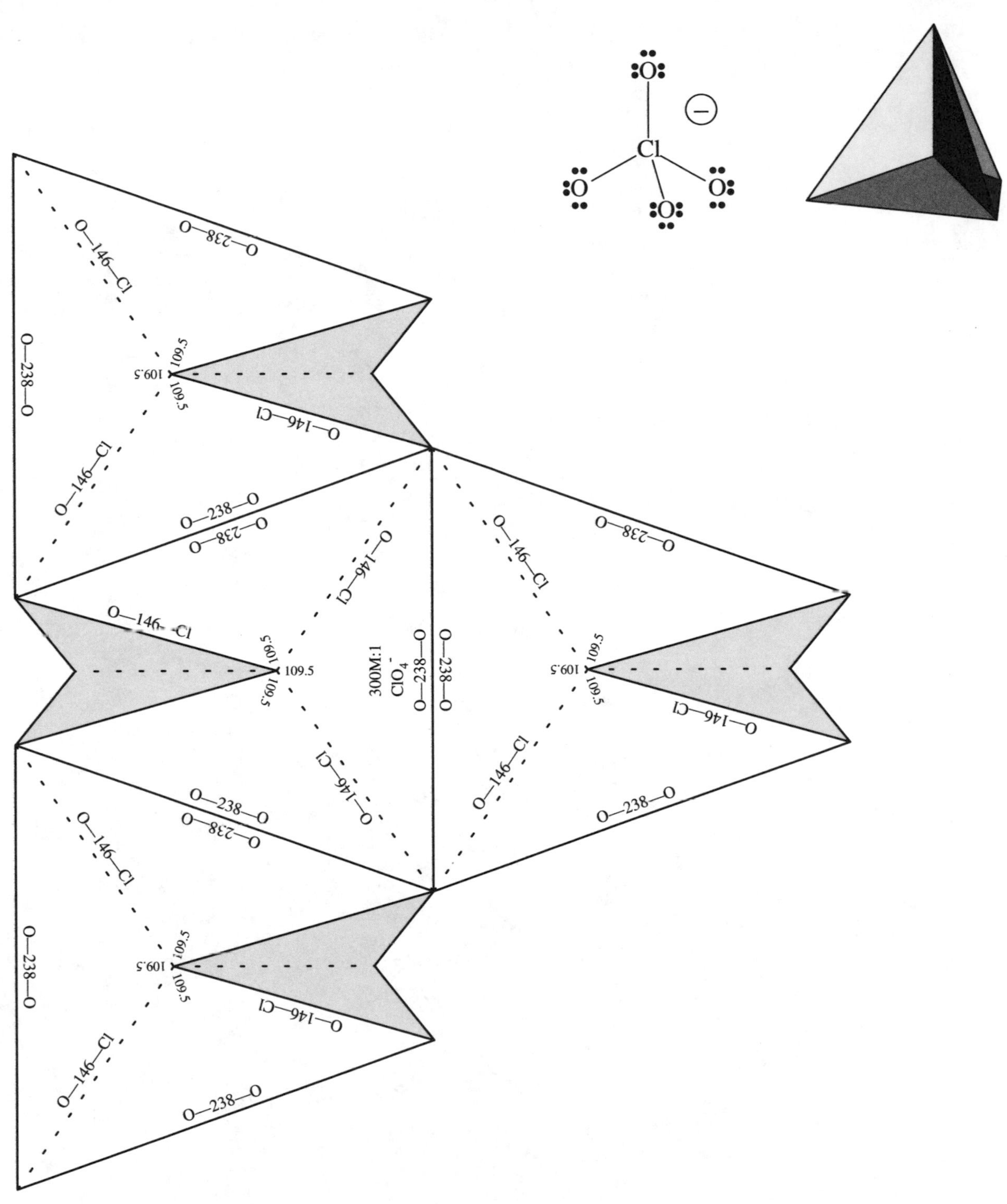

Question to think about:

In ClO_3^-, the Cl–O distances are 157 pm and the O–Cl–O angle is 106.7°. Is this what you would expect?

periodate ion (in $NaIO_4$) IO_4^-

shape: tetrahedral units: pm scale: 300,000,000:1

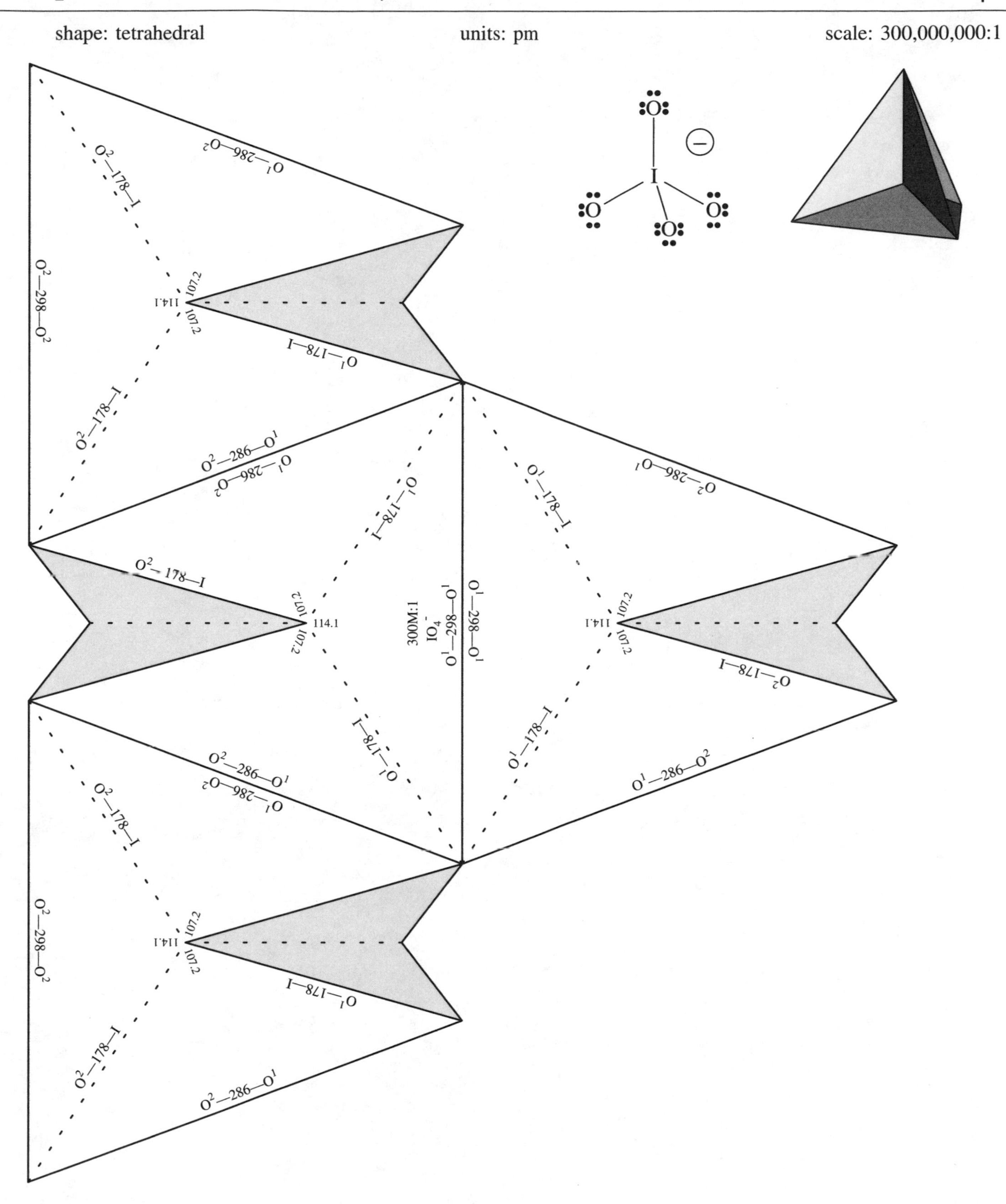

Question to think about:

Why aren't the angles in IO_4^- all 109.47°?

xenon tetroxide XeO_4

shape: tetrahedral units: pm scale: 300,000,000:1

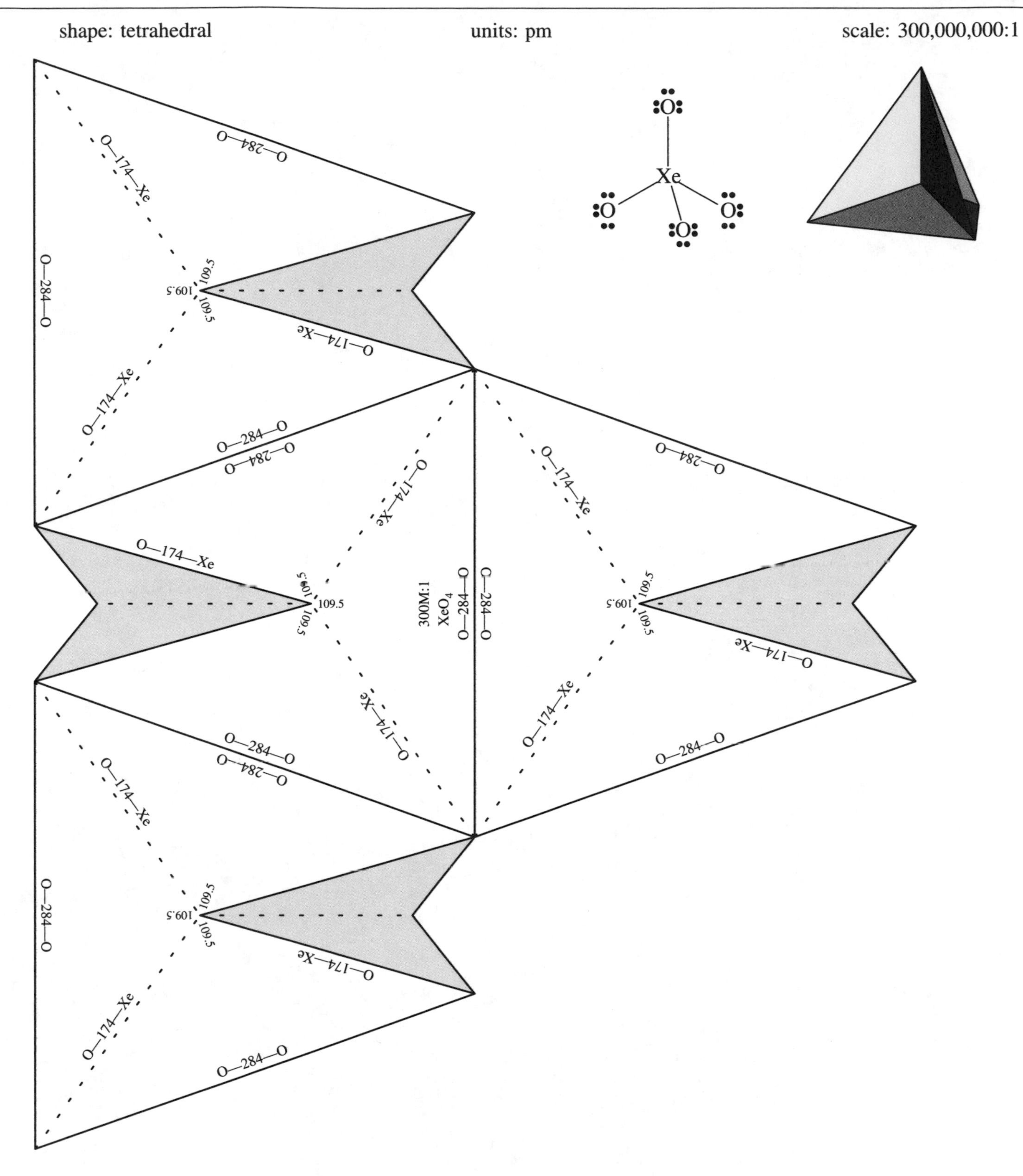

Question to think about:

How are IO_4^- and XeO_4 related? Is 174 pm reasonable for the Xe–O distance in XeO_4?

PART 2

Advanced Shapes

In Part 1, we focused on the two most common molecular shapes in nature, the trigonal pyramid and the tetrahedron. In this section, you will find four new shapes: **seesaw**, **trigonal bipyramid**, **square pyramid**, and **octahedron**:

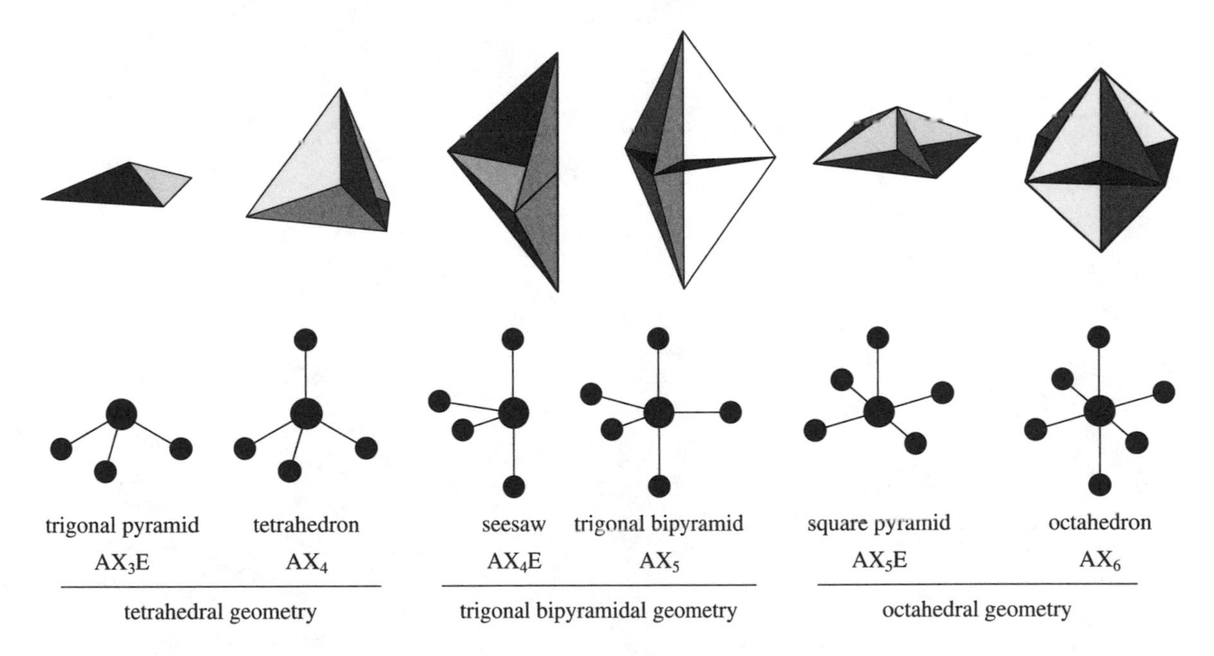

As can be seen, these six shapes fall into three distinct *geometries*, based on the number of overall "electron pairs" around the central atom. For example, both AX_3E and AX_4 have four electron pairs. The name of the geometry is taken from the shape having atoms at all positions. It can be very interesting to compare these structures among themselves and with those given in Part 1. In contrast to Part 1, no questions are included here. That shouldn't stop you from asking your own, though. Many of the same trends seen in Part 1 are seen here as well. Among the interesting comparisons are:

- **seesaw-shaped** SF_4 with SeF_4, SiF_4, and SF_6
- **trigonal bipyramid** PF_5 with SOF_4, PF_3, and PF_6^-
- **square pyramids** BrF_5 with $XeOF_4$ and BrF_3 (which is T-shaped), and $XeOF_4$ with XeF_4 (which is square planar)
- **octahedron** SF_6 with PF_6^-, SF_4, and SF_2

See if you can predict or rationalize these comparisons. What trends are apparent?

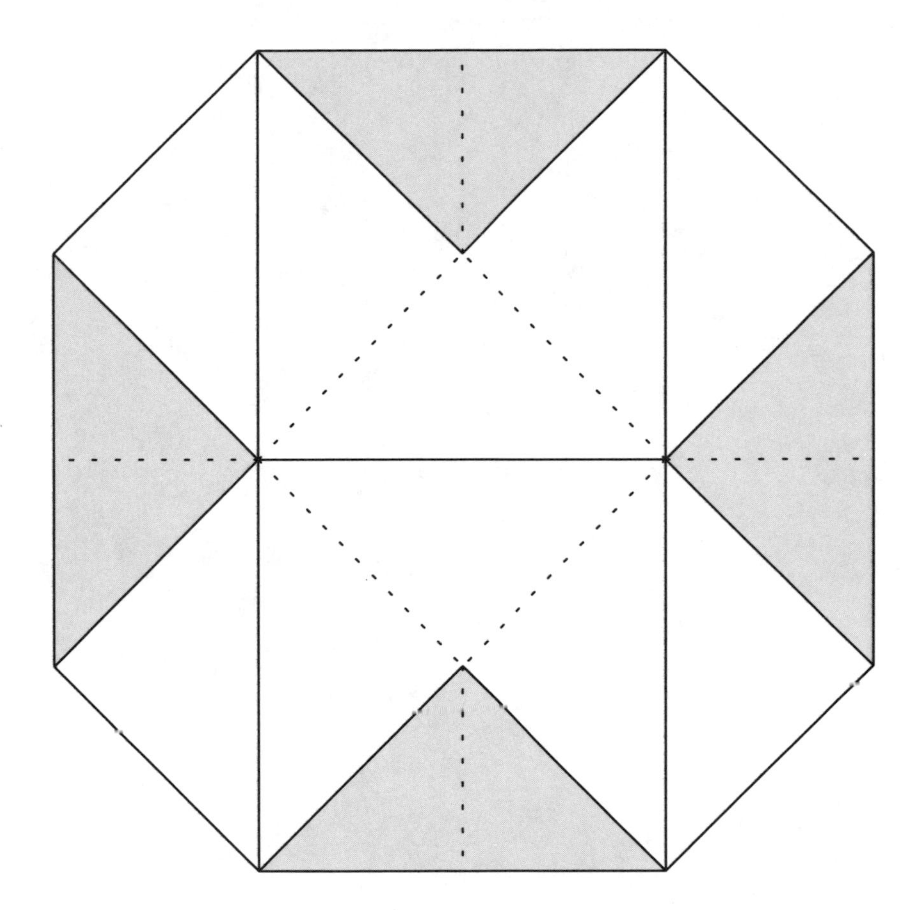

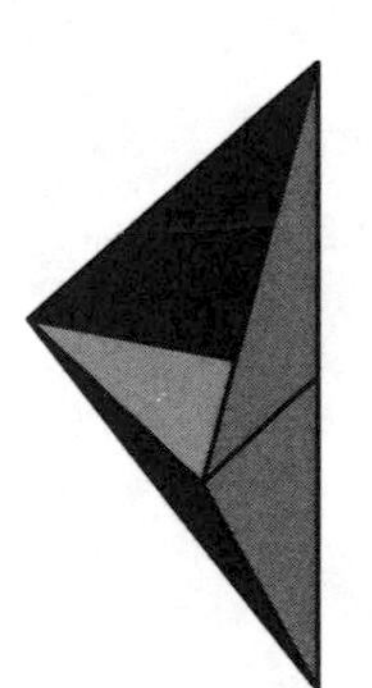

Seesaw-shaped molecules are much less common than trigonal pyramidal or tetrahedral molecules. They owe their shape to the presence of an additional pair of electrons associated with the central atom.

AX_4E molecules consist of a central atom surrounded by four outer atoms. Two of these outer atoms are almost directly across the molecule from each other (at the positions of the seats of a seesaw) and are called *axial*. The other two are positioned in the *equatorial plane*, under the fulcrum point of the seesaw. The lone pair of electrons fills out the third equatorial site.

The two models in this section show some representative aspects. Note, for example, that the axial distances are larger than the equatorial ones. In addition, the central atom is located slightly *outside* the region containing the outer atoms. Could this be due to the "size" of the lone pair, or is this the effect of the lone pair taking on more than its fair share of the available *s* character?

sulfur tetrafluoride SF_4

shape: seesaw units: pm scale: 240,000,000:1

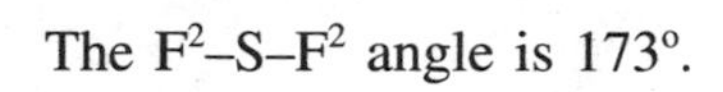

The F^2–S–F^2 angle is 173°.

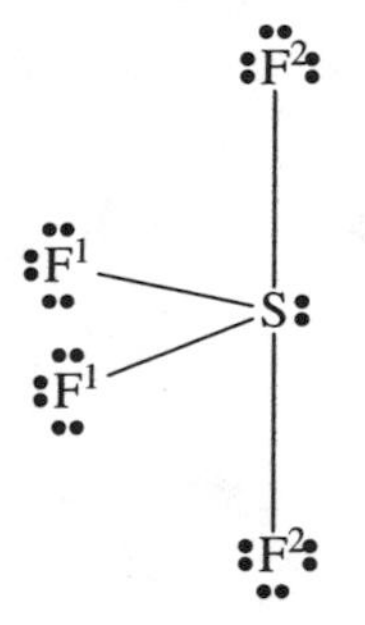

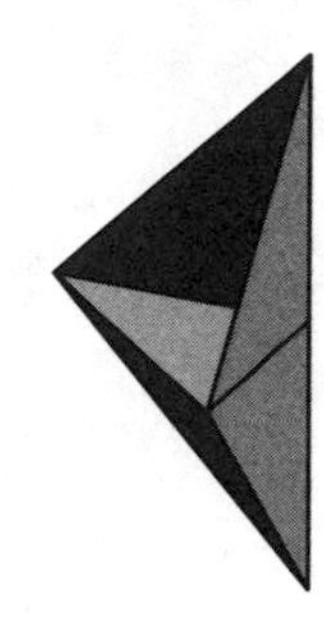

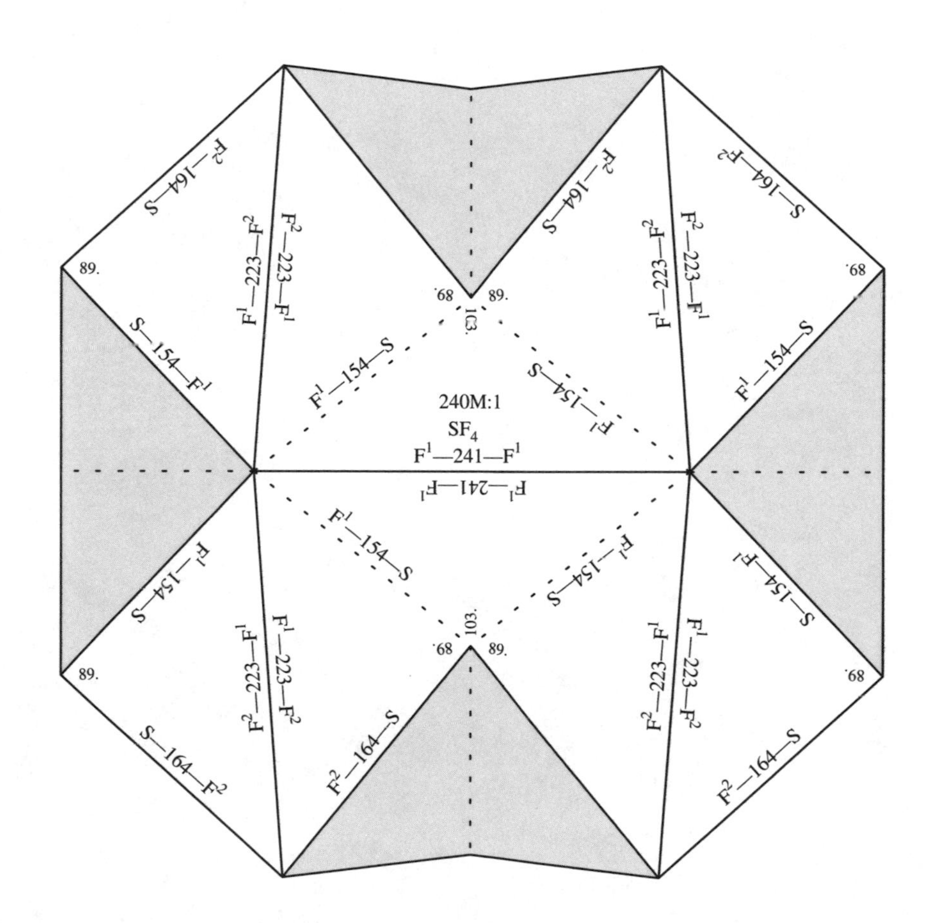

selenium tetrafluoride SeF_4

shape: seesaw   units: pm   scale: 240,000,000:1

The F^2–Se–F^2 angle is 169.2°.

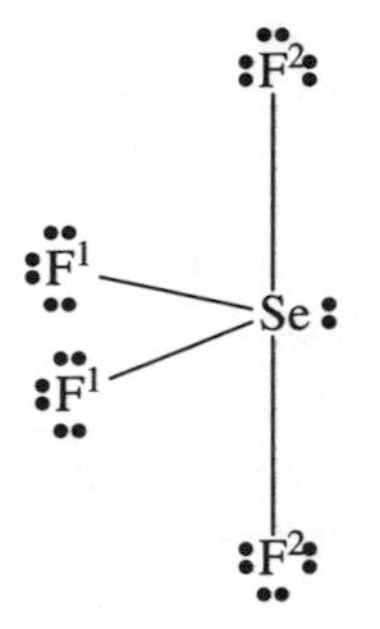

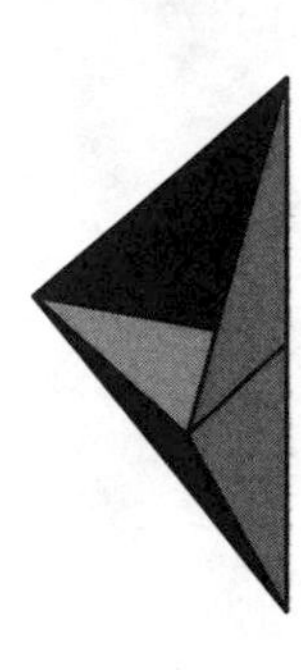

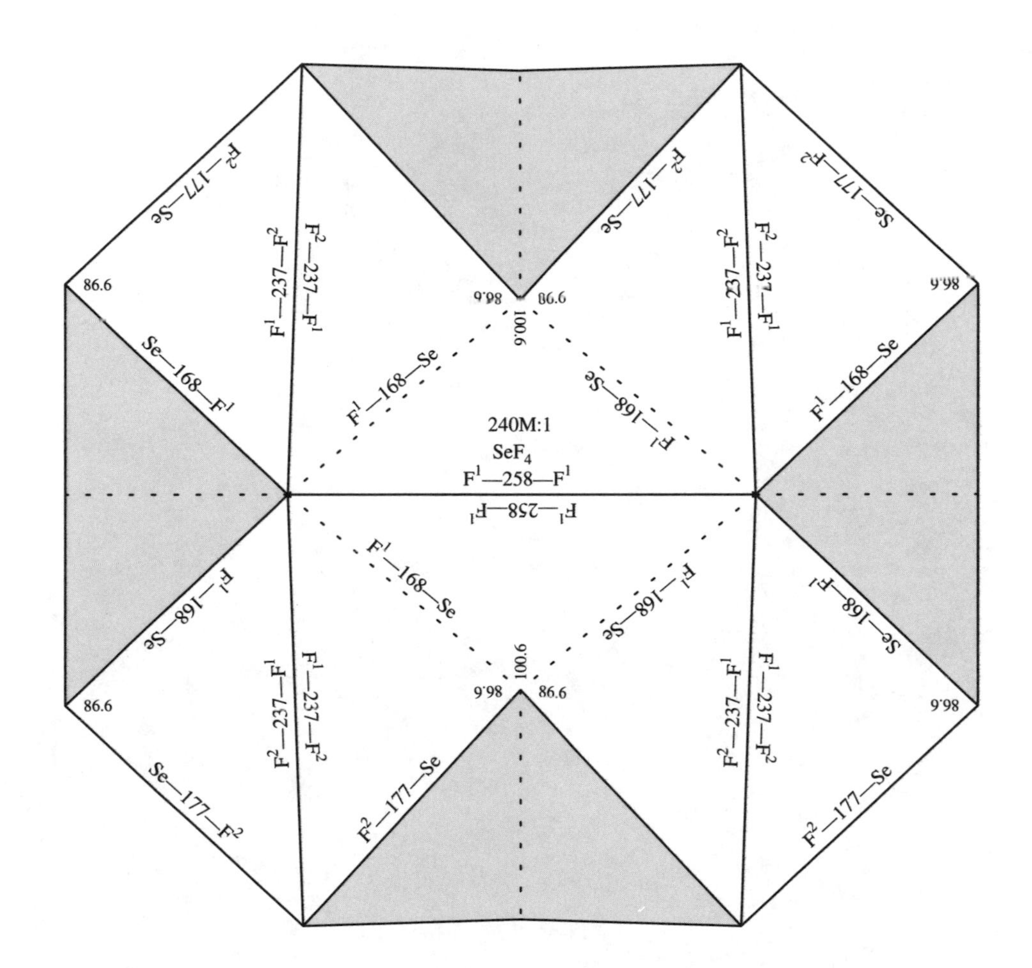

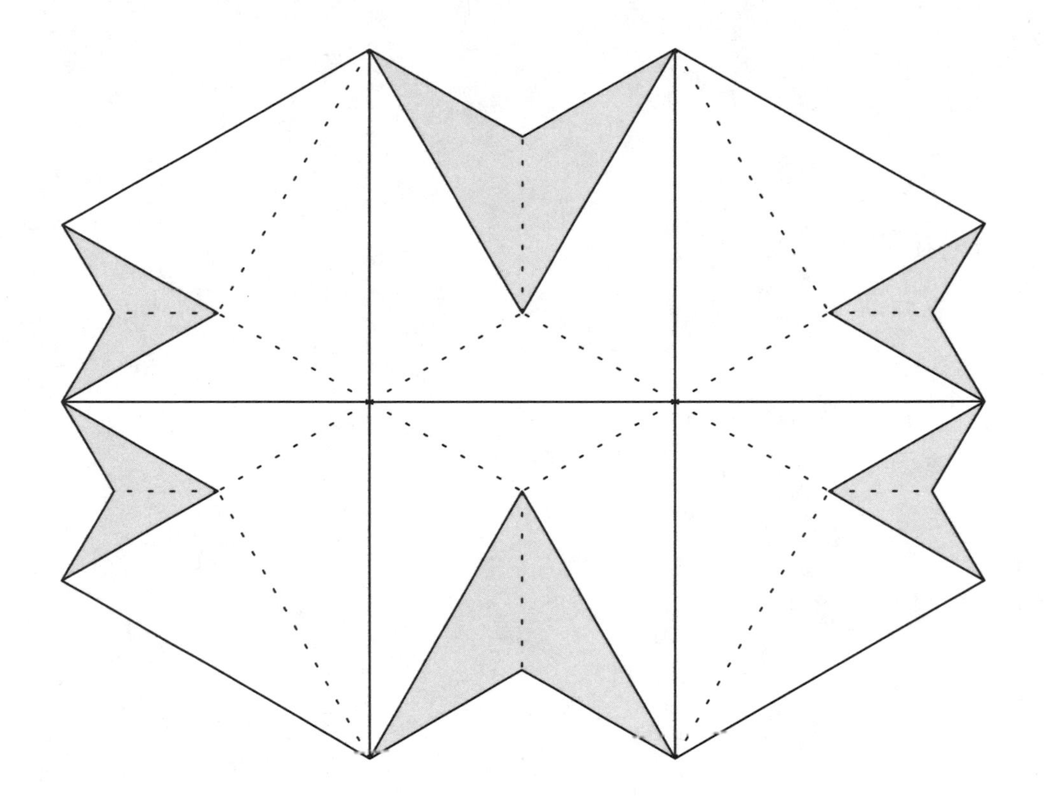

Trigonal bipyramidal molecules consist of five outer atoms distributed around a central atom. Three of these outer atoms are in the *equatorial* plane. The other two atoms are directly opposite each other and are called *axial.*

An idealized AX_5 model is shown above and consists of six faces. In this particular pattern, each X_{eq}–A–X_{eq} angle is 120°; each X_{eq}–A–X_{ax} angle is 90°.

The X_{eq}–A–X_{eq} angles in actual molecules vary. For example, in SOF_4 (page 91), F_{eq}–S–F_{eq} is 110° and O–S–F_{eq} is 125°. In addition, the angles between the axial X groups and the equatorial plane are not always 90°, although they are always very close to this value.

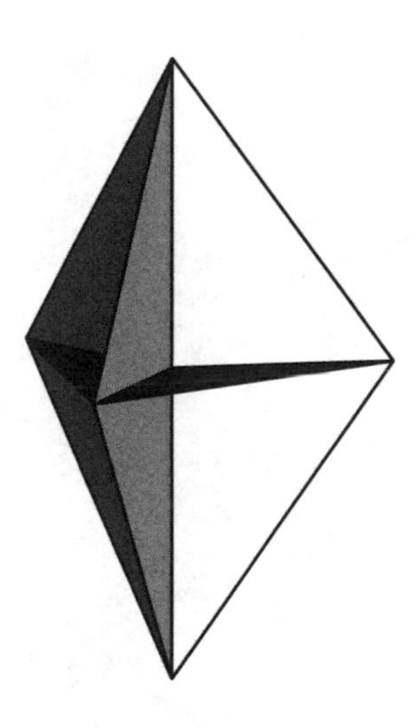

phosphorus pentafluoride PF_5

shape: trigonal bipyramidal | units: pm | scale: 240,000,000:1

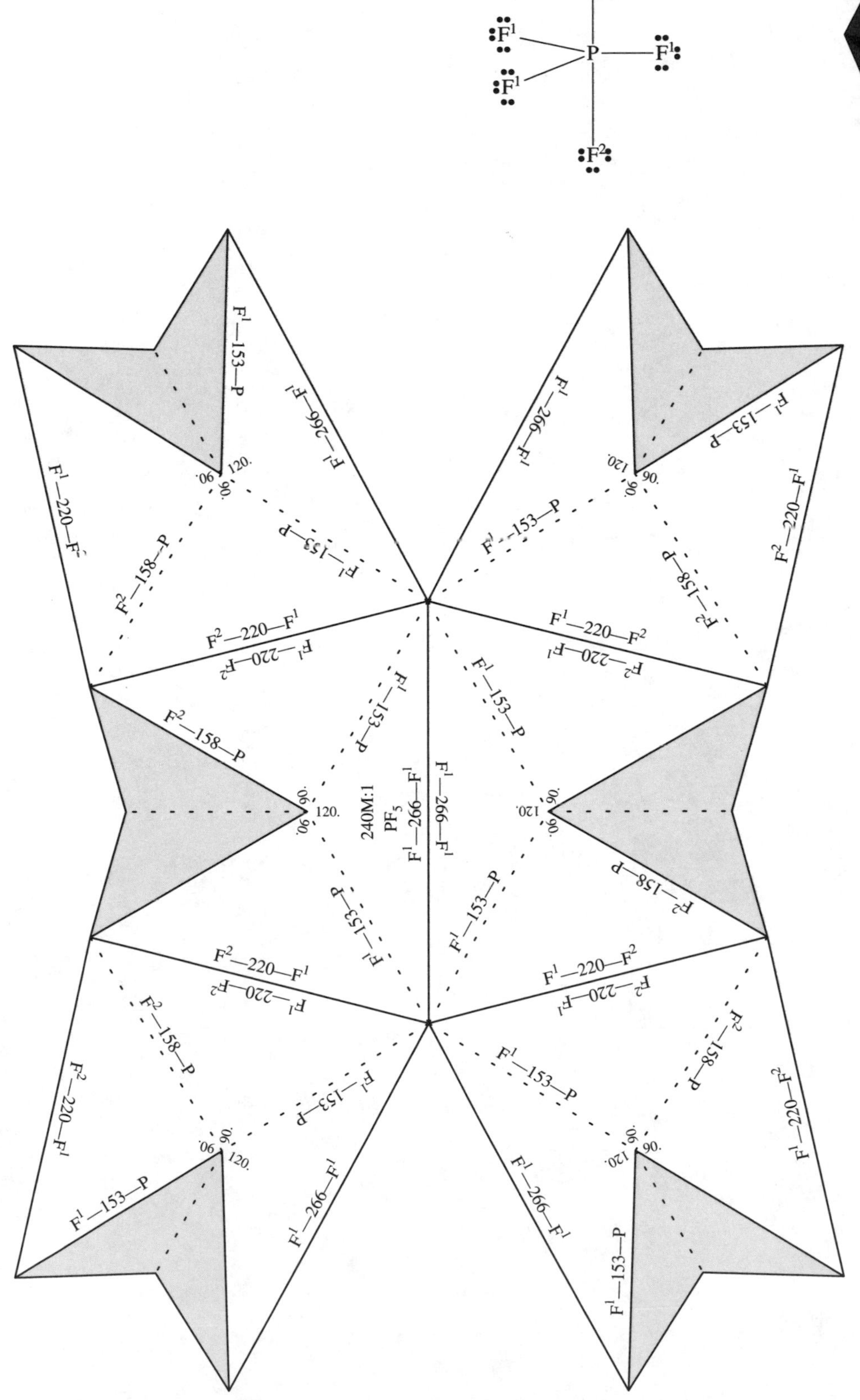

sulfur oxide tetrafluoride

shape: trigonal bipyramidal | units: pm | scale: 240,000,000:1

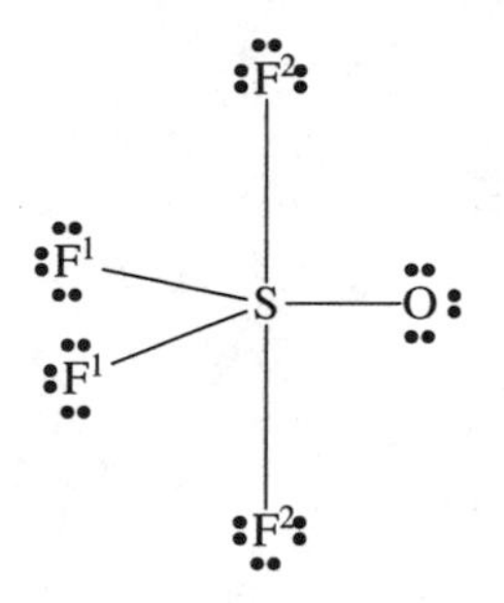

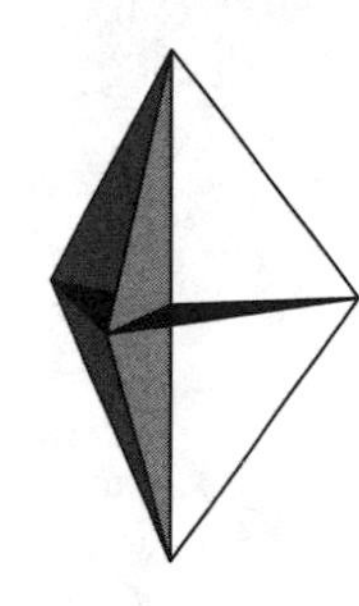

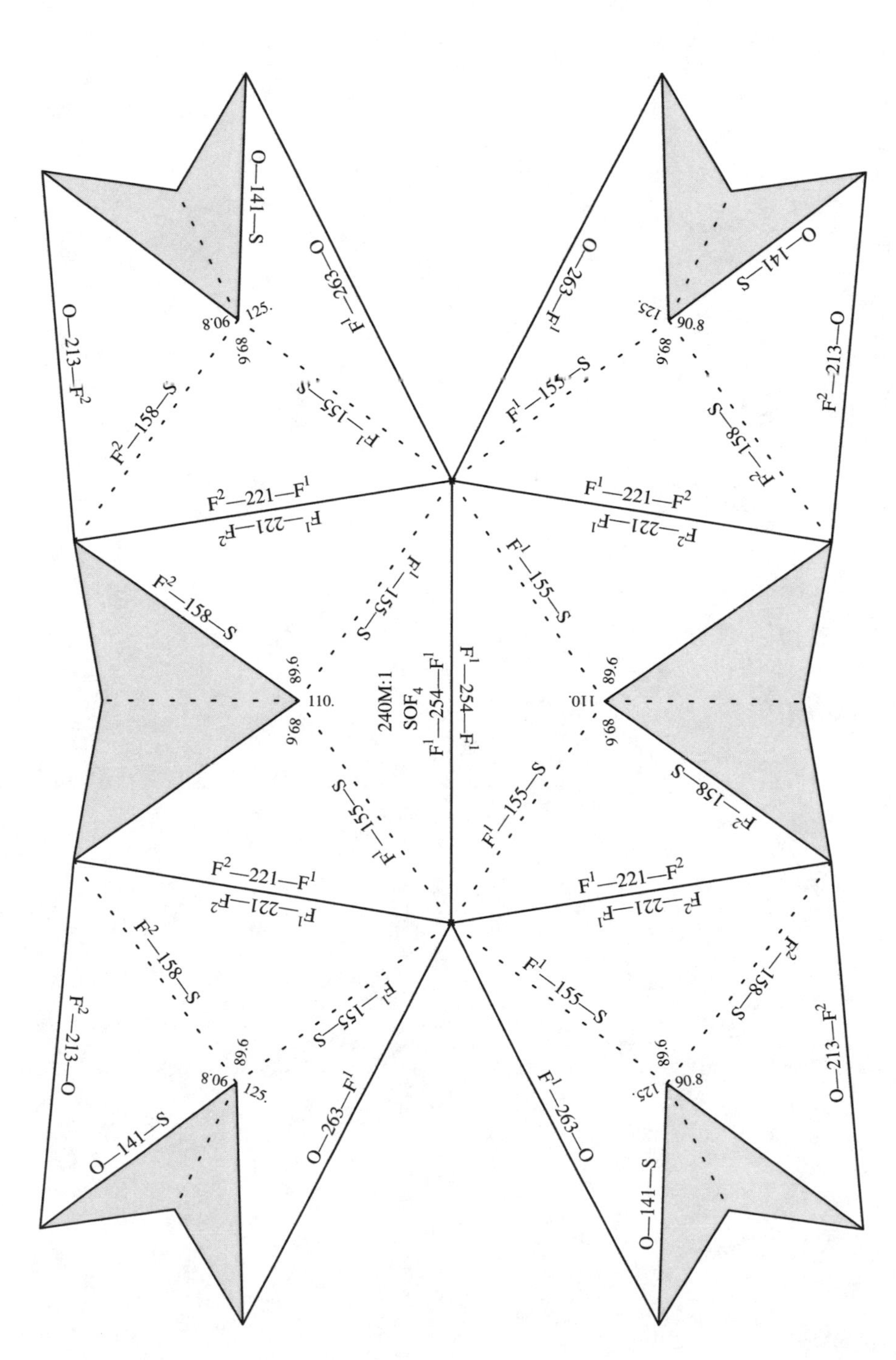

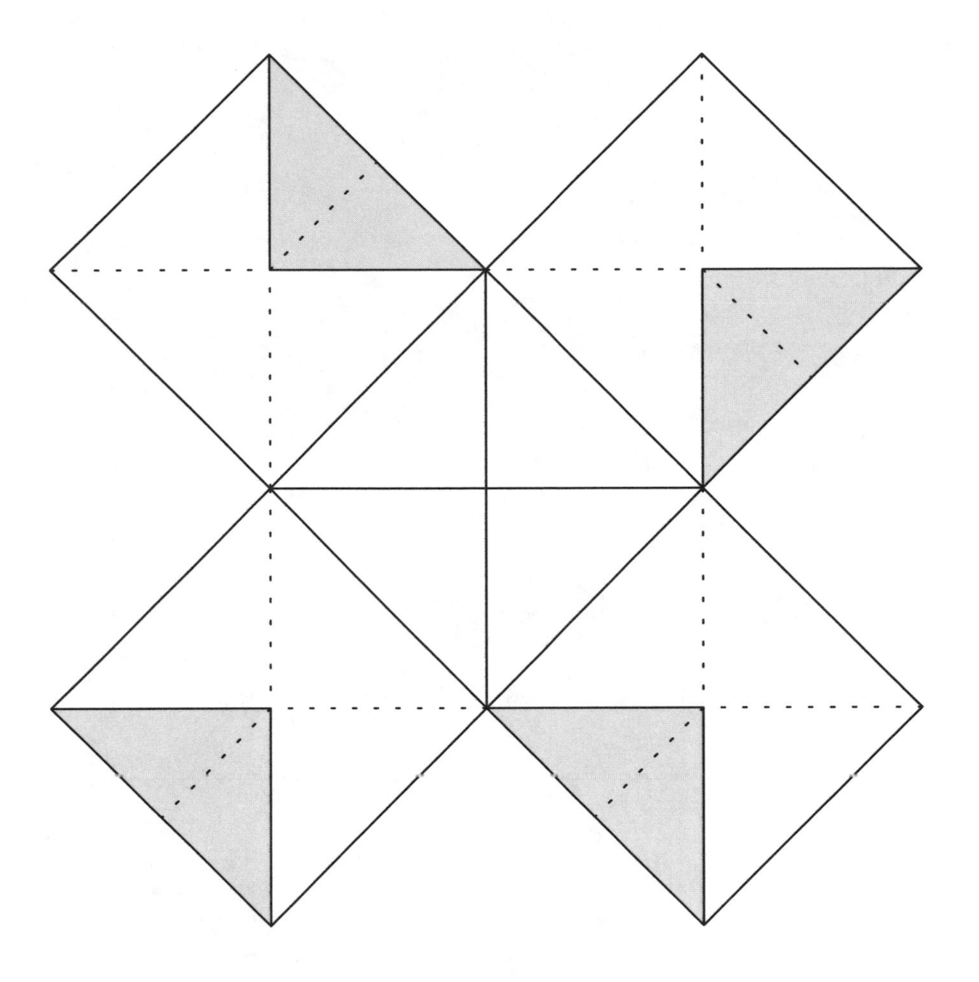

Square pyramidal molecules involve five outer atoms and a lone pair of electrons around a central atom, AX_5E. One of the outer atoms is at the top of the pyramid in the *apical* position. The other four atoms form the base of the pyramid and thus are called *basal* atoms. The lone pair of electrons fills out the sixth site.

In all real AX_5E cases, the apical A–X bond distance is less than the basal A–X bond distance. In addition, among the AX_5E molecules, the apical-center-basal atom angle varies considerably. For example, it is 85° in BrF_5 (page 95) and 92° in $XeOF_4$ (page 97). Note that in BrF_5 the central atom is *below* the basal plane (presumably because of lone-pair repulsion of the basal bond electrons). Even with this variation, the basal-central-basal atom angles range only from 89.5° to 89.9°.

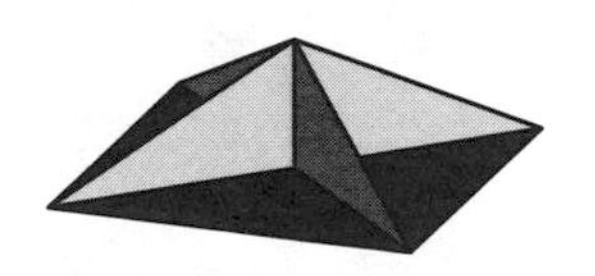

bromine pentafluoride BrF_5

shape: square pyramidal units: pm scale: 240,000,000:1

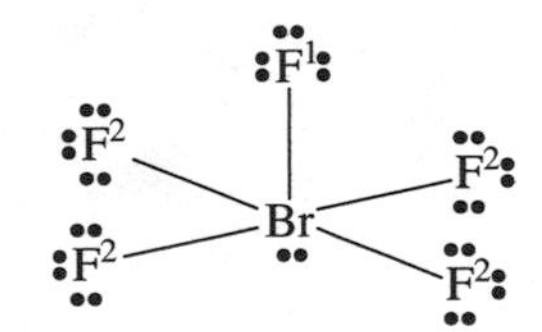

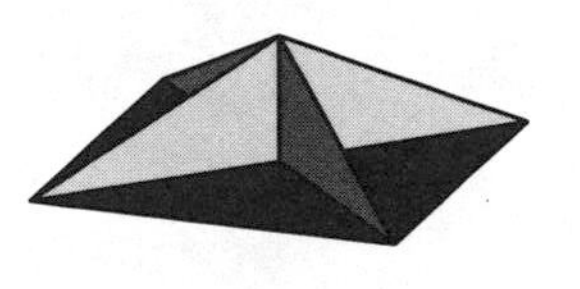

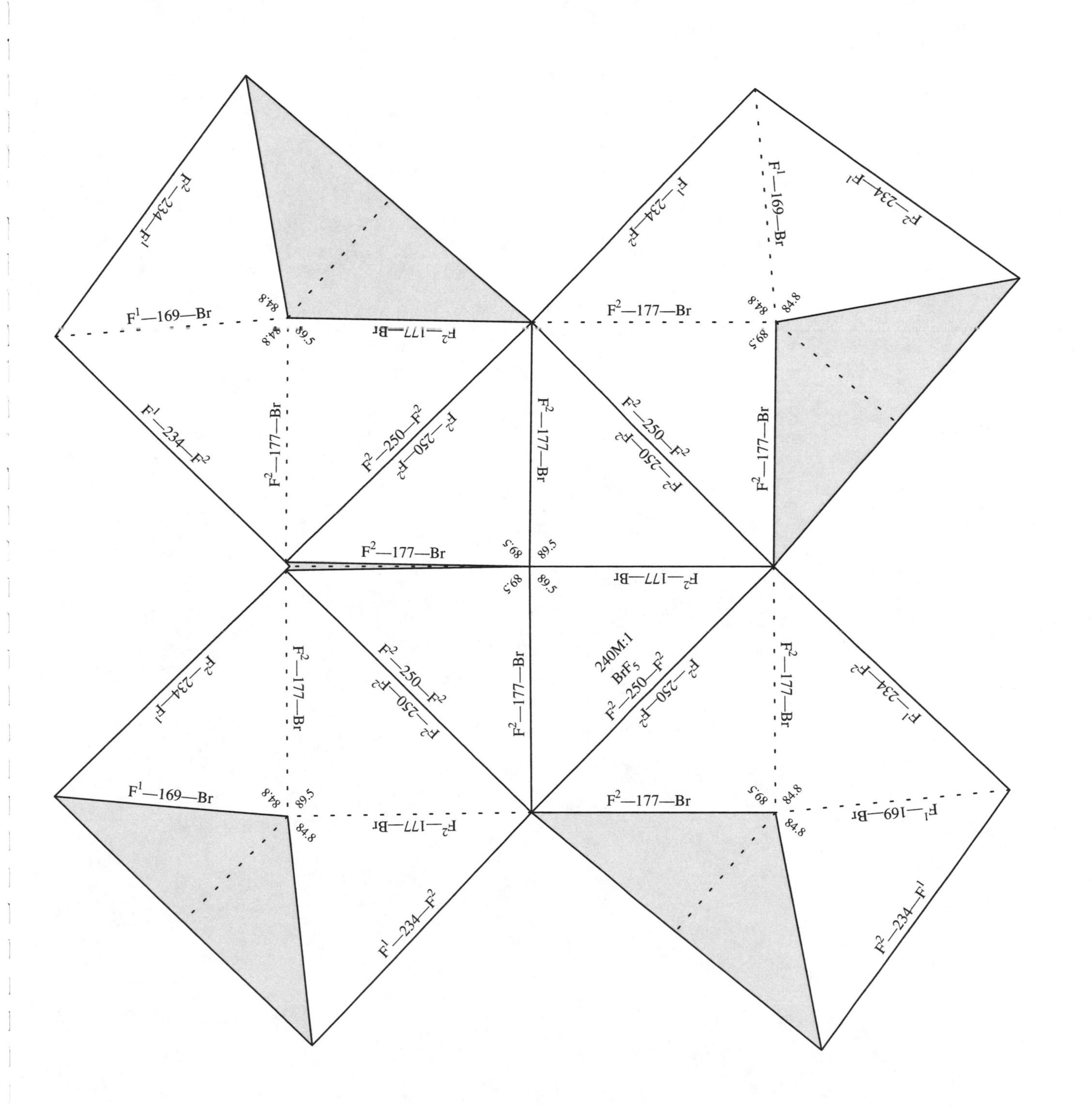

xenon oxytetrafluoride $XeOF_4$

shape: square pyramidal units: pm scale: 240,000,000:1

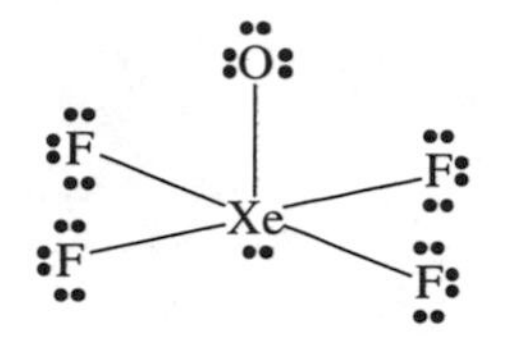

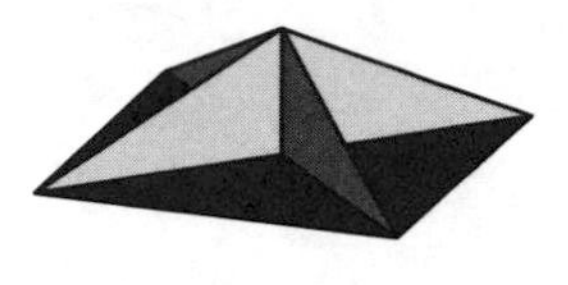

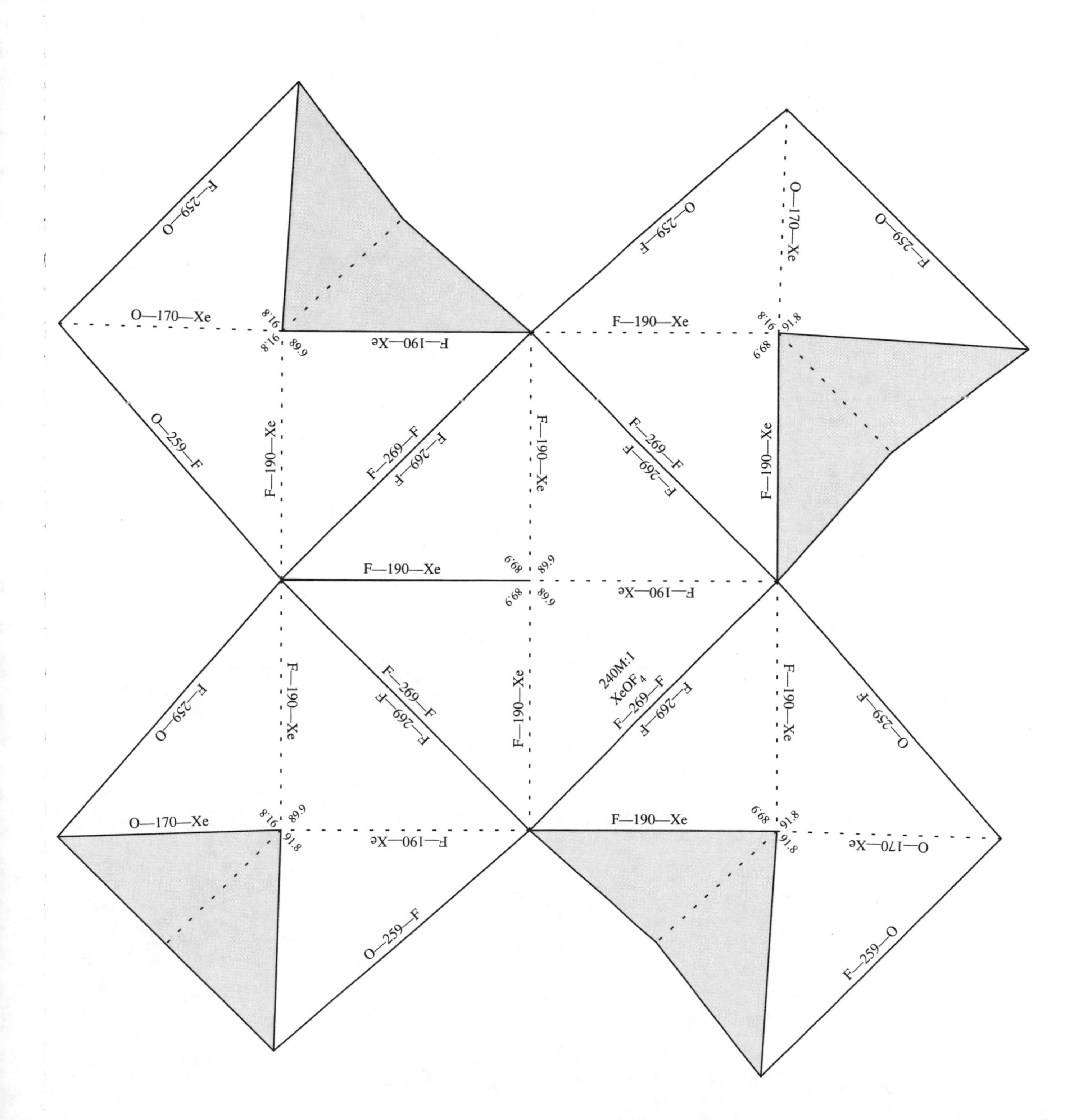

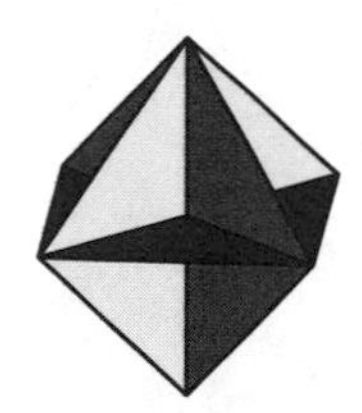

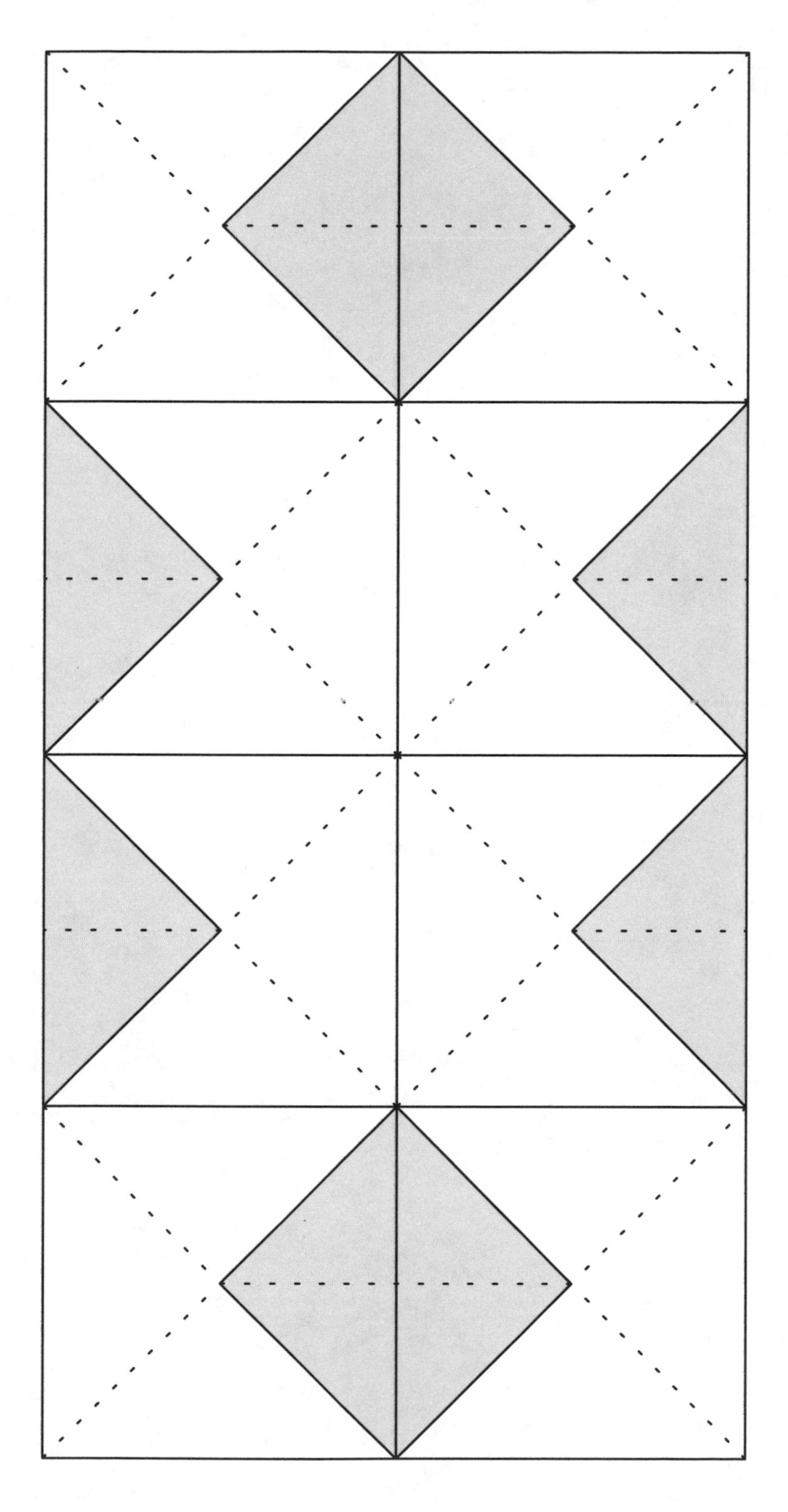

Octahedral molecules consist of a central atom surrounded by six outer atoms. In octahedral molecules, groups across from each other are referred to as *trans*. Groups in adjacent positions are called *cis*.

The basic pattern shown here, used for SF_6 (page 101), produces a model with all of the distances the same and all of the angles 90°. Although this highly idealized geometry is found for many molecules, especially in the gas phase, many octahedral molecules and ions are not so regular. For example, in PF_6^- as found in $NaPF_6{\cdot}H_2O$ (page 103), four of the P–F distances are 158 pm, while two are 173 pm.

The two models in this section show a significant decrease in size as one proceeds across a row of the periodic table keeping the number of electrons the same. The decrease in X–A distance fits in with a larger series involving several known ions not modeled here: AlF_6^{-3} (181 pm), SiF_6^{-2} (171 pm), PF_6^- (158 pm), and SF_6 (156 pm).

sulfur hexafluoride

SF_6

shape: octahedral

units: pm

scale: 240,000,000:1

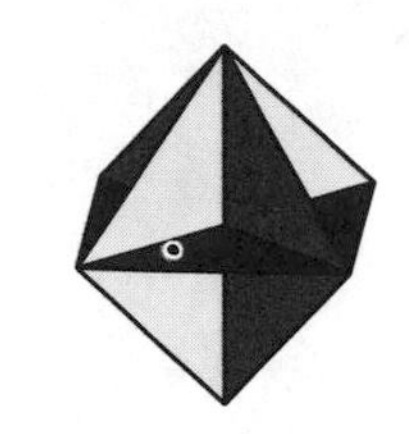

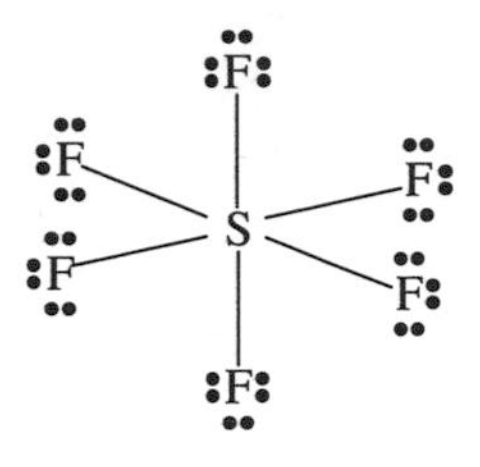

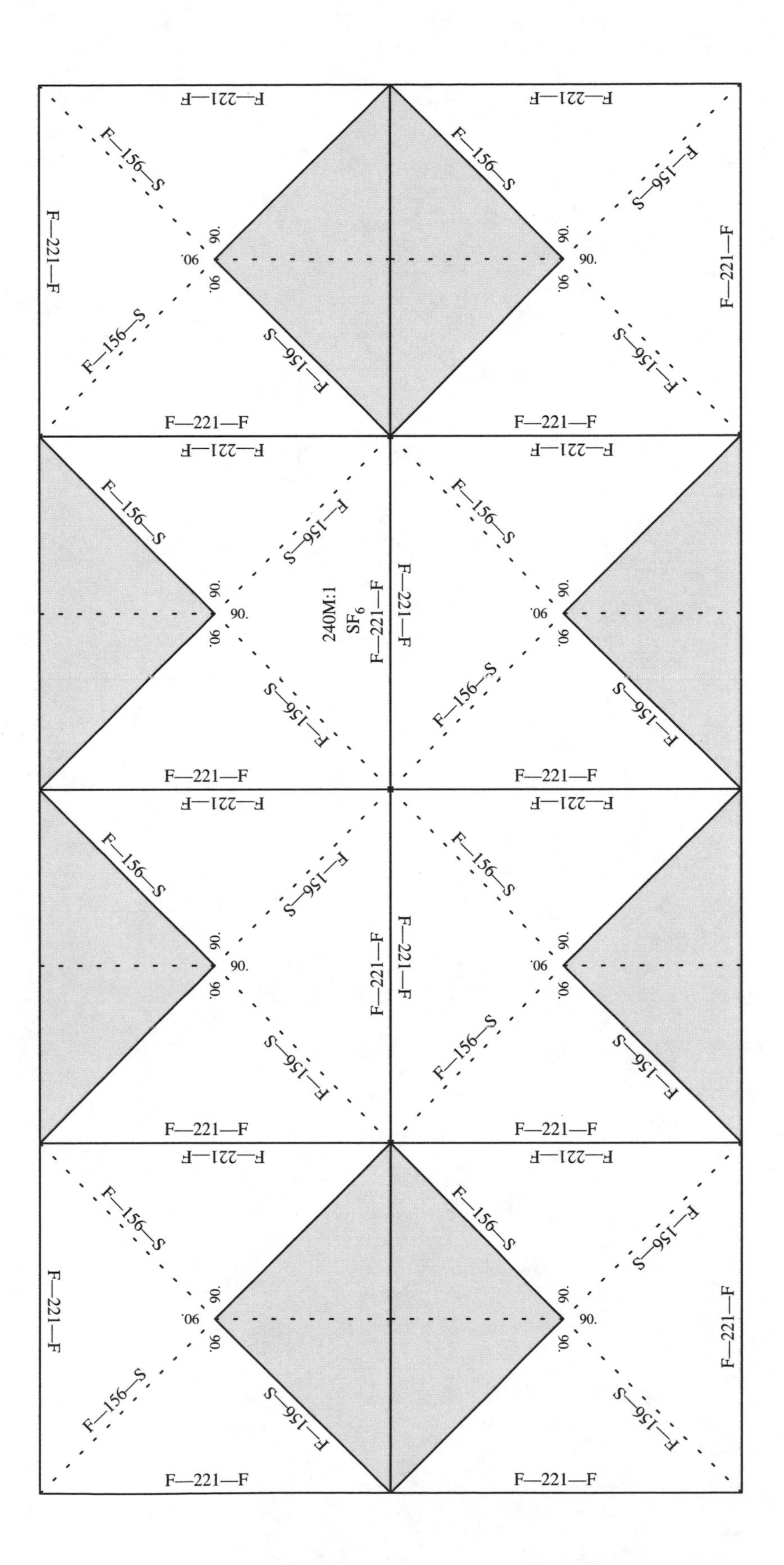

phosphorus hexafluoride ion (in $NaPF_6 \cdot H_2O$) PF_6^-

shape: octahedral units: pm scale: 240,000,000:1

Don't fold center lines. Tape the two halves back-to-back.

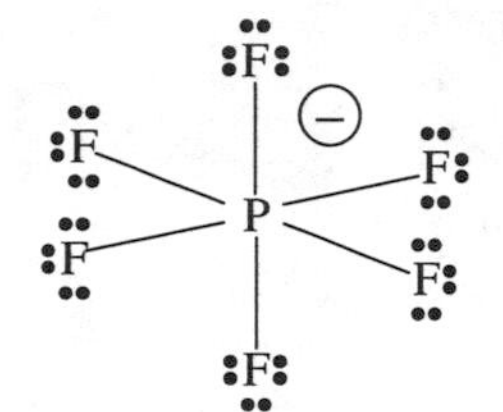

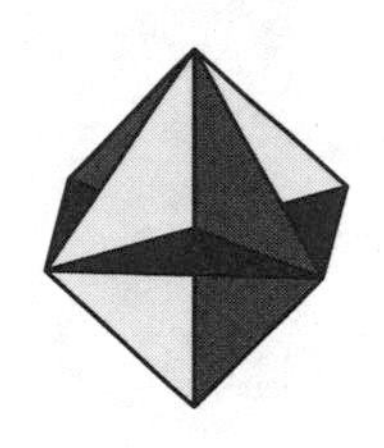

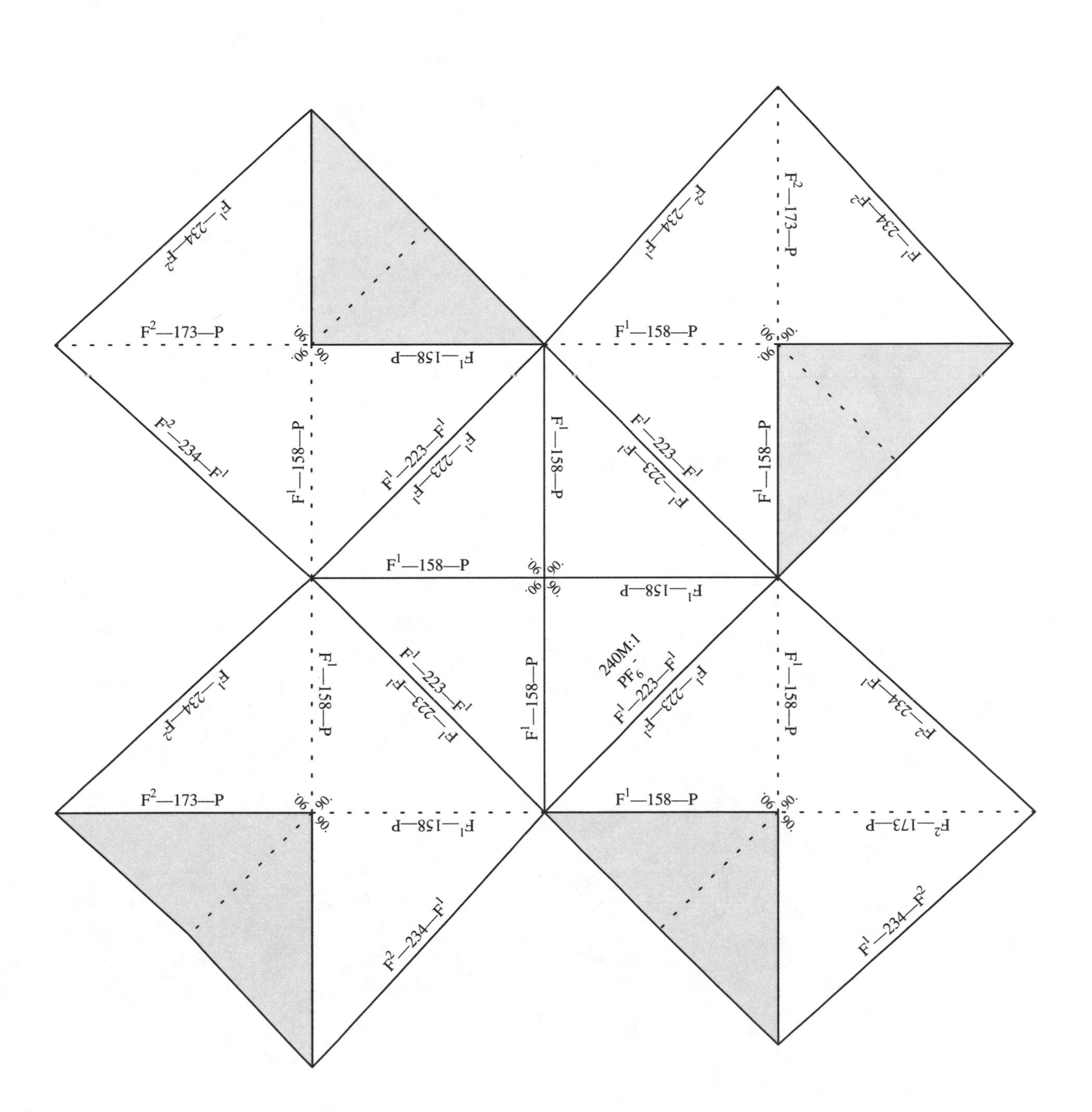

phosphorus hexafluoride ion (in $NaPF_6 \cdot H_2O$) (second copy)

PF_6^-

shape: octahedral | units: pm | scale: 240,000,000:1

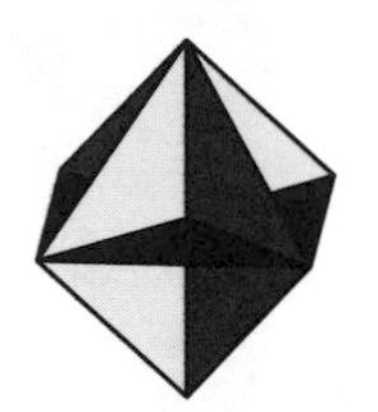

Don't fold center lines.

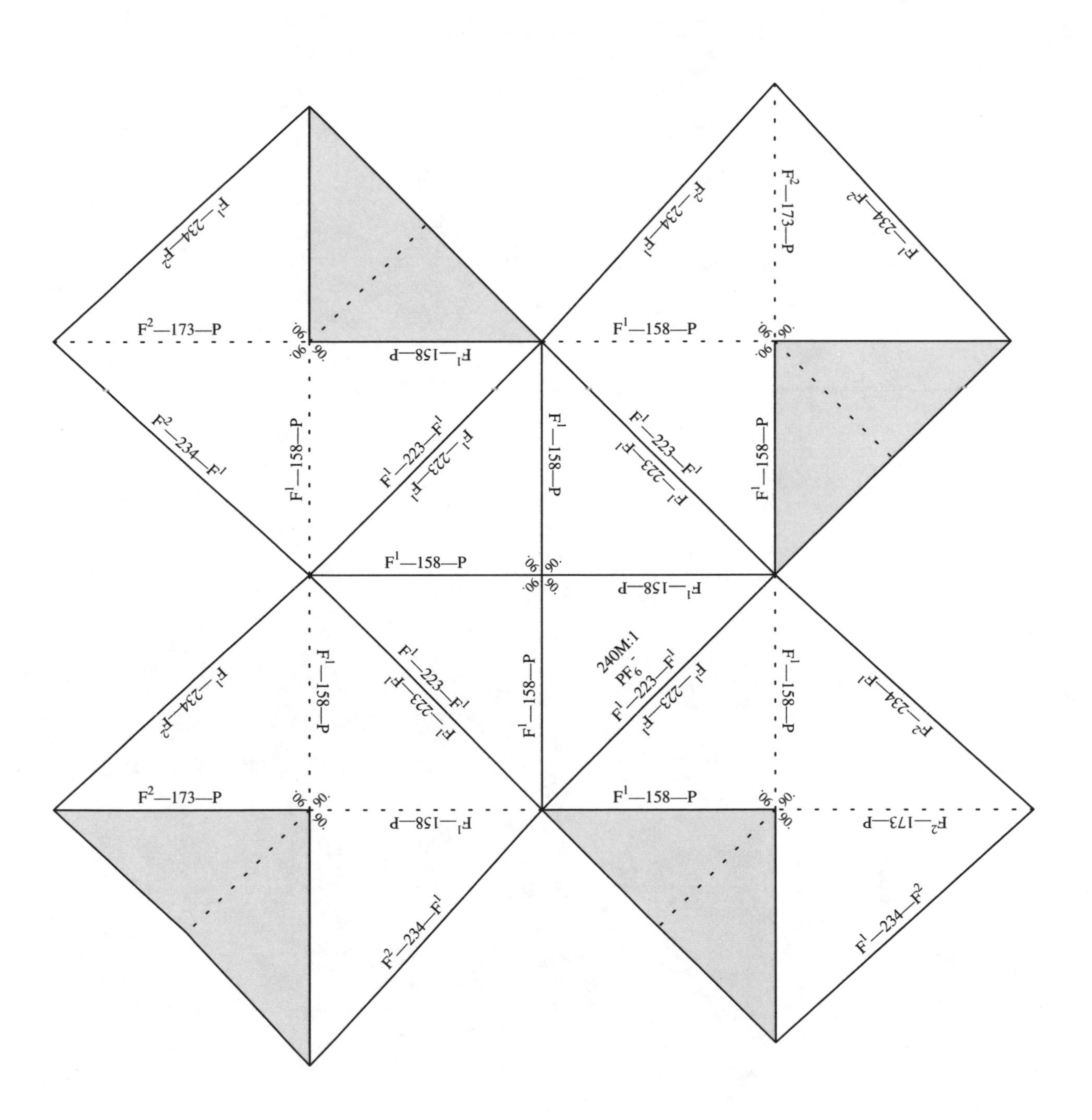

PART 3

Beyond Octahedra

In this section, the last to introduce new shapes, you will find six of the least understood molecular shapes in chemistry:

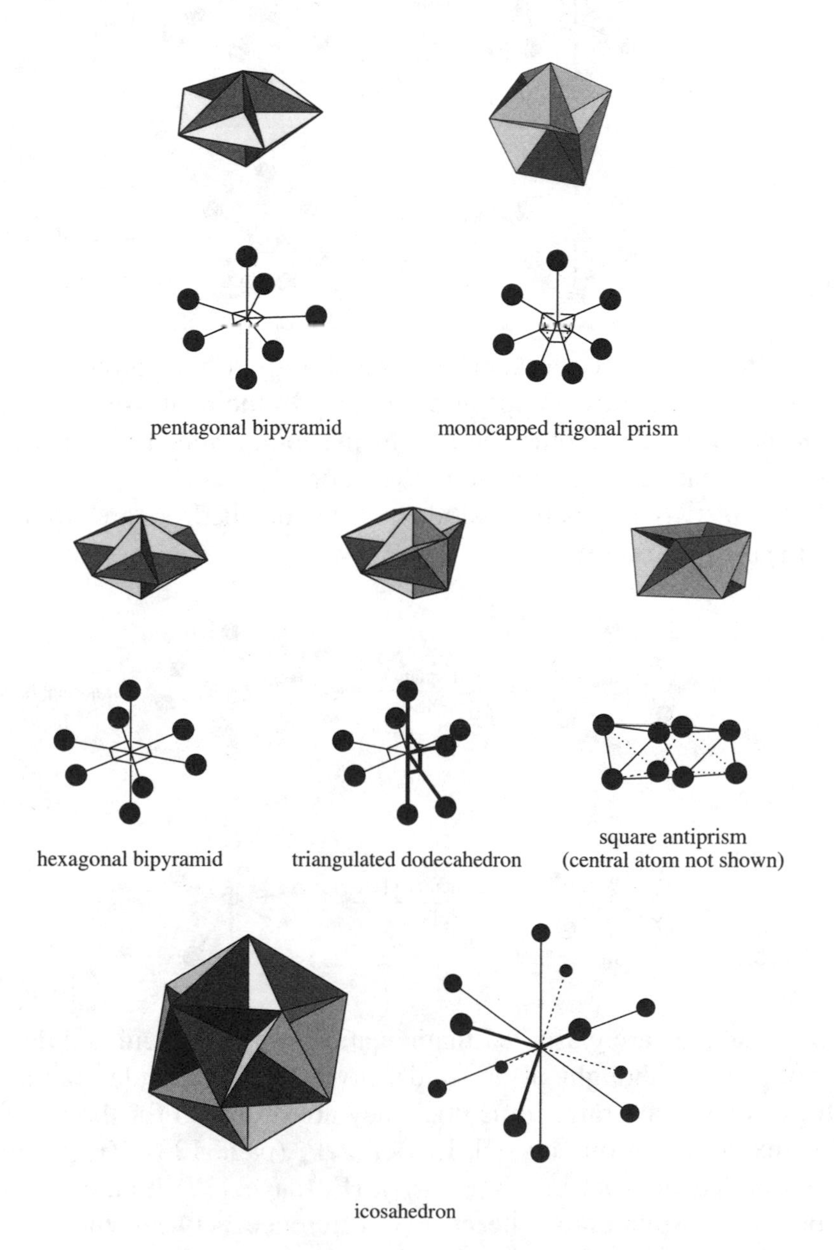

Much of the difficulty in describing these shapes, I think, arises from the inherent difficulty we have in seeing three dimensions on paper. If you have trouble visualizing one or another of these shapes (as I do), *make a model*! I recommend making one for a real molecule, so you can see some of the subtle distortions from ideality that come into play in nature.

At first glance, these six shapes may seem completely unrelated. Nonetheless, they can all be derived mathematically relatively easily from a regular octahedron. For example, if one were to imagine splitting one of the six positions of an octahedron to make a seven-coordinate system, there are two basic ways to proceed:

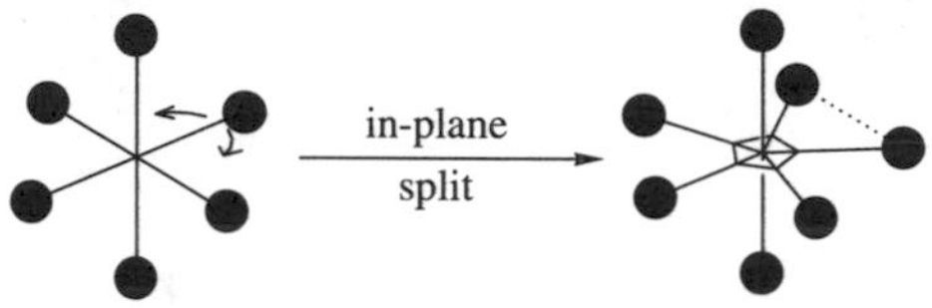

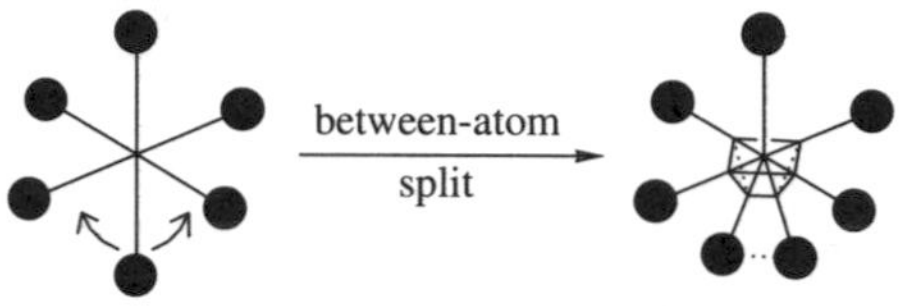

The only difference between the pentagonal bipyramid and the monocapped trigonal prism, then, is how the position is split. In the pentagonal bipyramid it is split *in the plane* of three other atoms. In the monocapped trigonal prism, it is split in a plane that runs *between* the other atoms.

Similarly, if two opposite positions of an octahedron are split, an eight-coordinate system is produced:

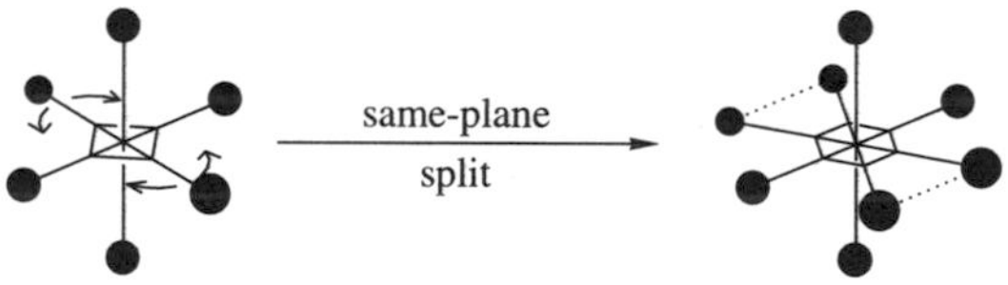

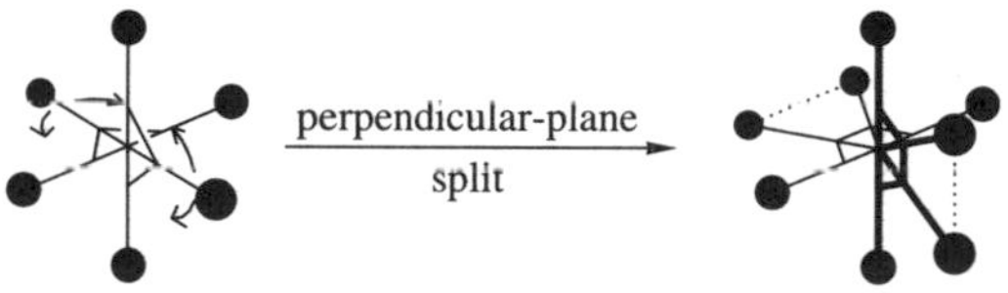

Both of these shapes are classified mathematically as "triangulated dodecahedra" because both, when thought of as solids, involve 12 triangular faces. Although both of these shapes are rare, sometimes they are observed for the same molecule or ion. For example, in one crystal, $Li_6BeF_4ZrF_8$ (page 123), ZrF_8^{-4} appears as a triangulated dodecahedron; in $[Cu(H_2O)_6]_2ZrF_8$ (page 127) it takes the shape of a square antiprism. Apparently, the energy difference between the two shapes is quite small. In fact, the triangulated dodecahedron and the square antiprism are simply "twisted" forms of each other. The only difference is a small twist of two different four-atom groups:

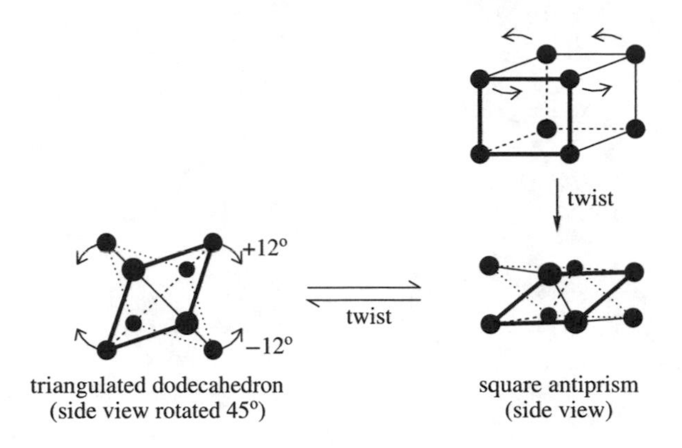

triangulated dodecahedron (side view rotated 45°)

square antiprism (side view)

Finally, the icosahedron arises from splitting *all six* of the positions, each pair of opposites splitting in a different plane:

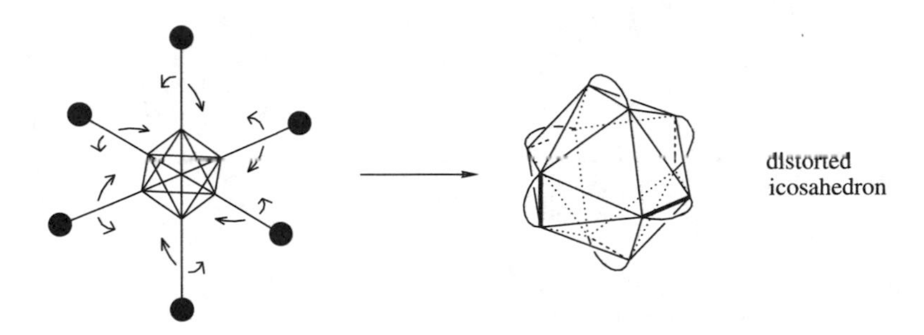

Depending upon how much "splitting" occurs, the icosahedron may be distorted. Nowhere is this distortion more obvious than in $Ce(NO_3)_6^{-3}$ (page 135), but it is also subtly present even in the icosahedral ion $B_{12}H_{12}^{-2}$ (page 151, Part 4).

If one carries this idea of splitting even further and splits all twelve of the icosahedral vertices *five* times (5 × 12 = 60) one arrives at a shape called the *truncated icosahedron*, an example of which is buckminsterfullerene, C_{60} (page 155).

heptafluorouranate(IV) ion (in K_3UF_7) UF_7^{-3}

shape: pentagonal bipyramidal units: pm scale: 180,000,000:1

This is the top half.

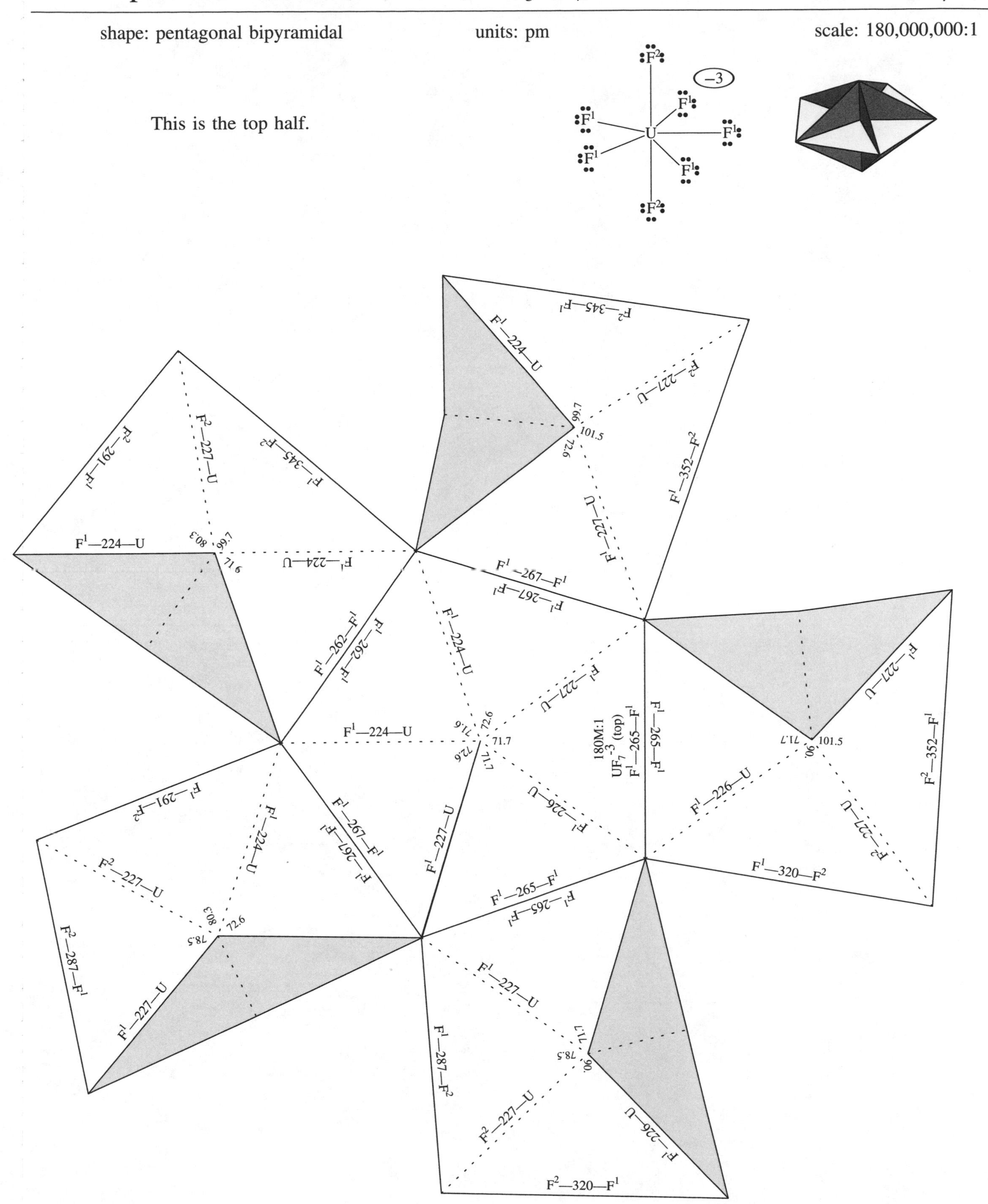

heptafluorouranate(IV) ion (bottom half) UF_7^{-3}

shape: pentagonal bipyramidal units: pm scale: 180,000,000:1

This is the bottom half. The central lines must be creased. Align the 262-pm edges on each half when combining.

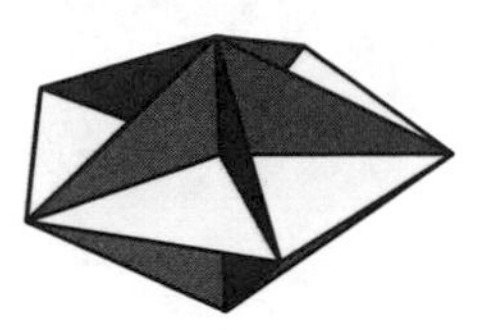

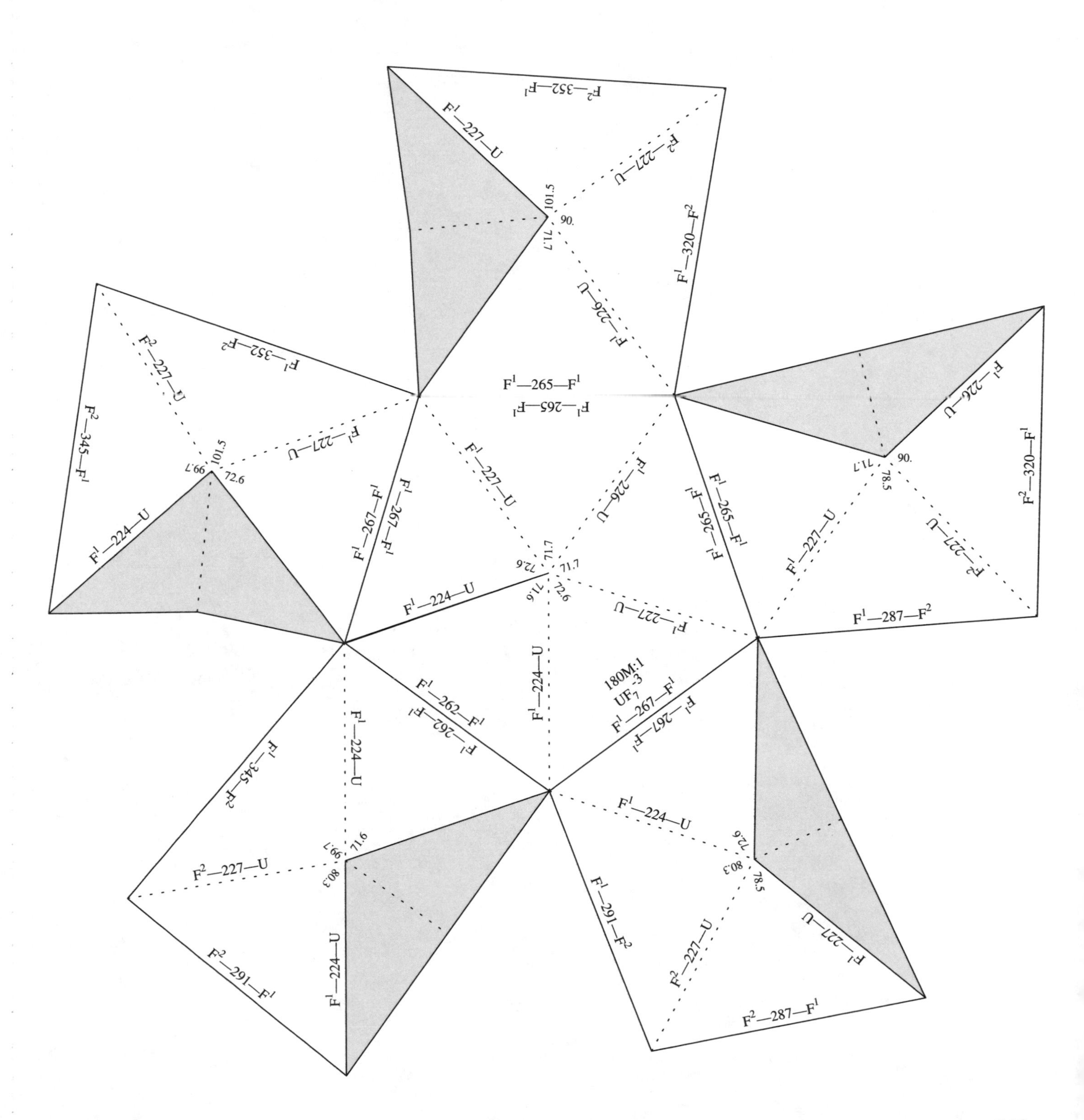

heptafluoroniobate(V) ion (in K_2NbF_7) NbF_7^{-2}

shape: monocapped trigonal prism units: pm scale: 180,000,000:1

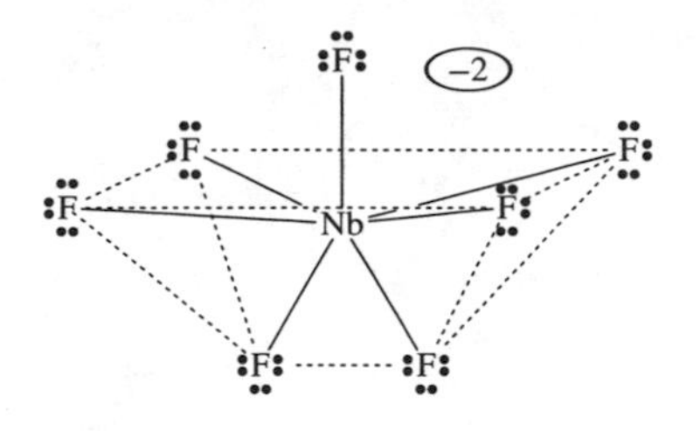

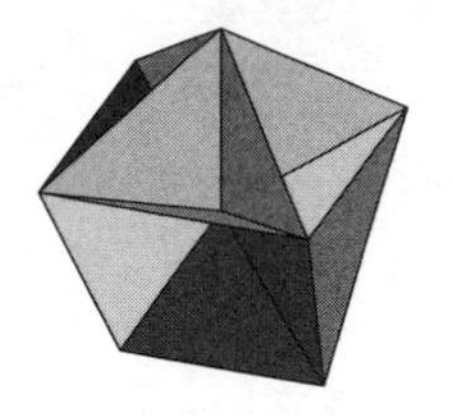

This is the prism.
F[1] and F[3] are at the bottom.

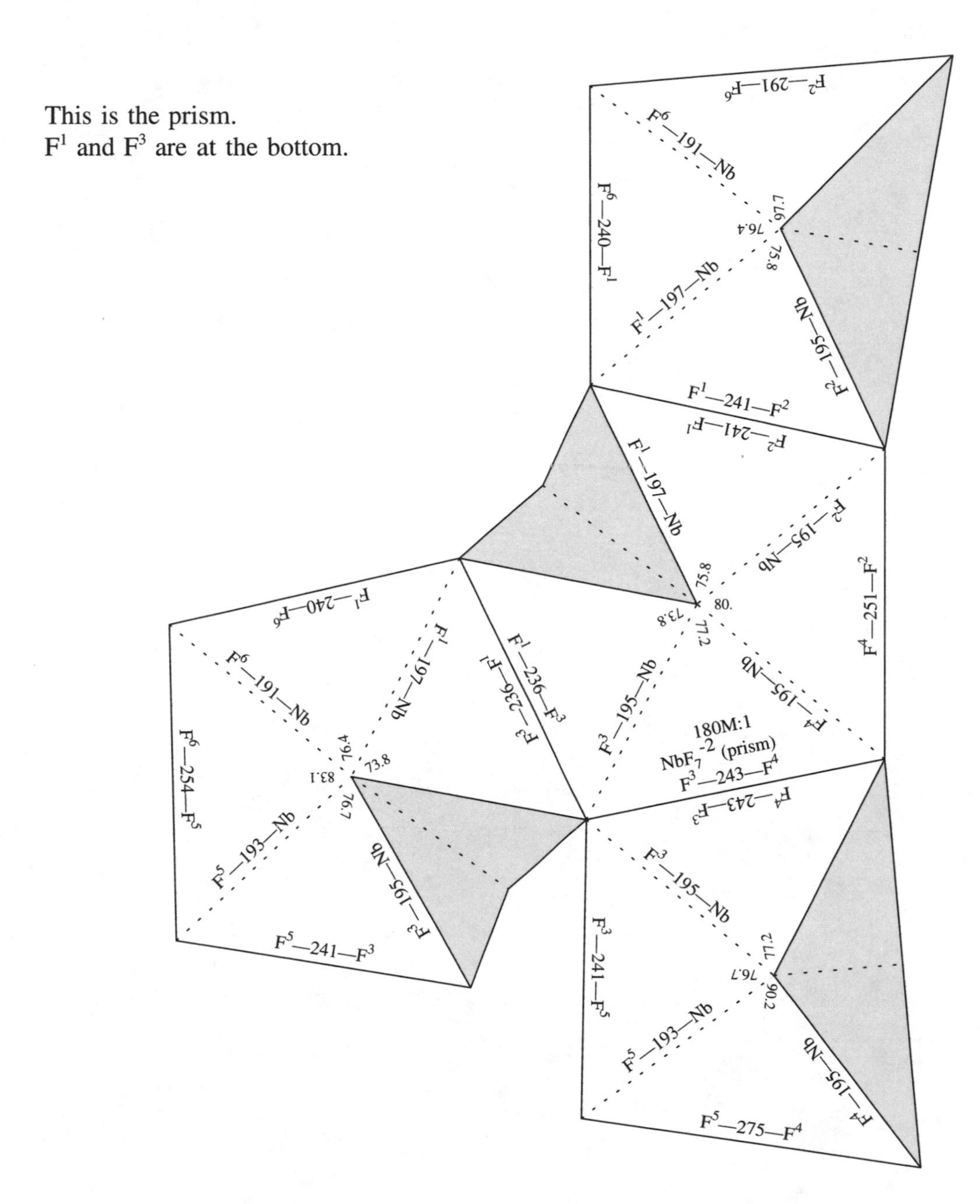

heptafluoroniobate(V) ion (cap part) NbF_7^{-2}

shape: monocapped trigonal prism units: pm scale: 180,000,000:1

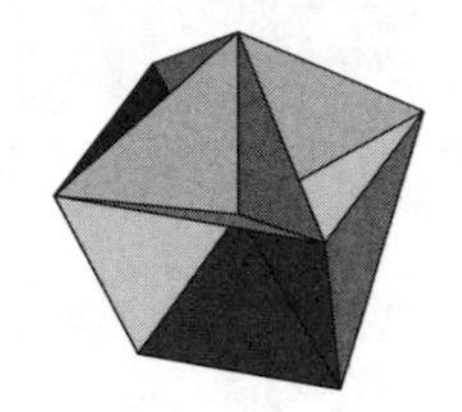

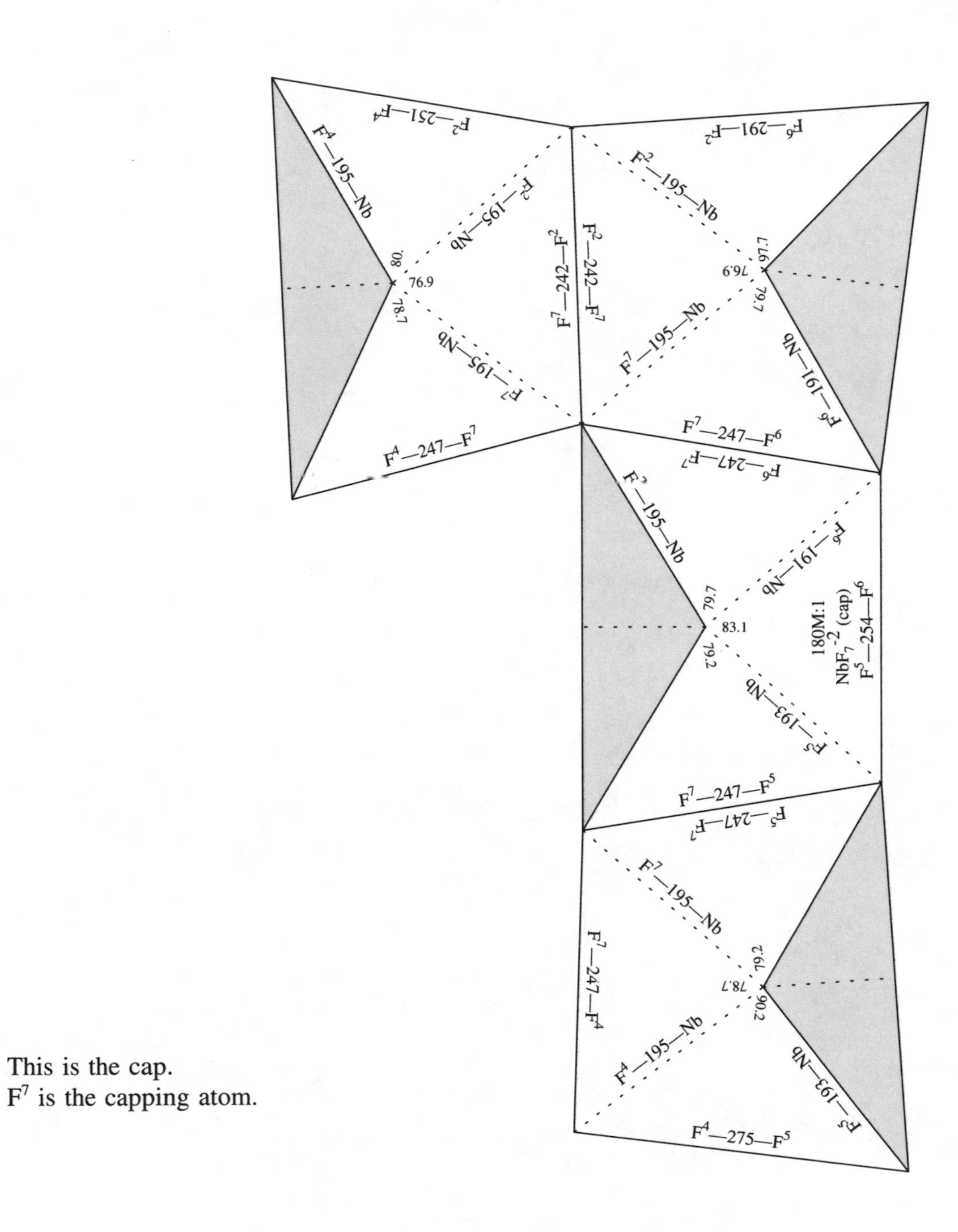

This is the cap.
F^7 is the capping atom.

uranyl nitrate ion (in $RbUO_2(NO_3)_3$) $UO_2(NO_3)_3^-$

shape: hexagonal bipyramid units: pm scale: 180,000,000:1

This is the front half. O^2 is axial. Fold the NO_3^- pattern in half and connect the folded edge to the main model across the short (216-pm) O^1–O^1 edge.

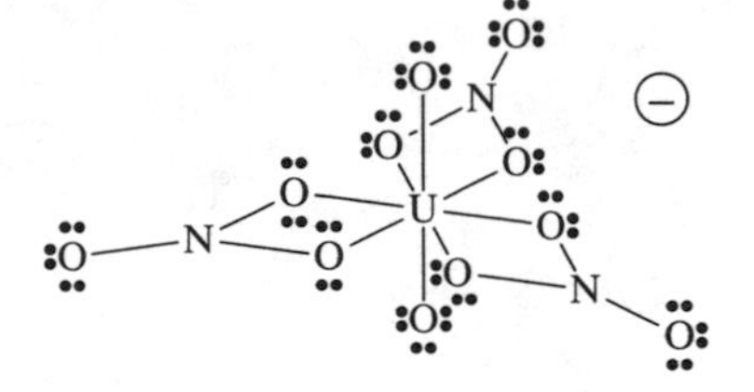

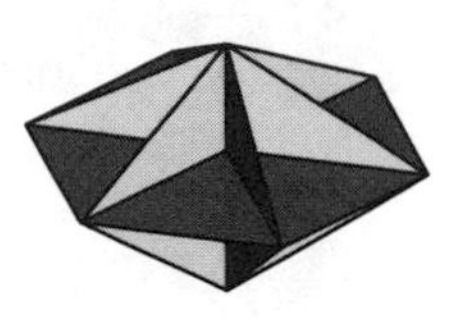

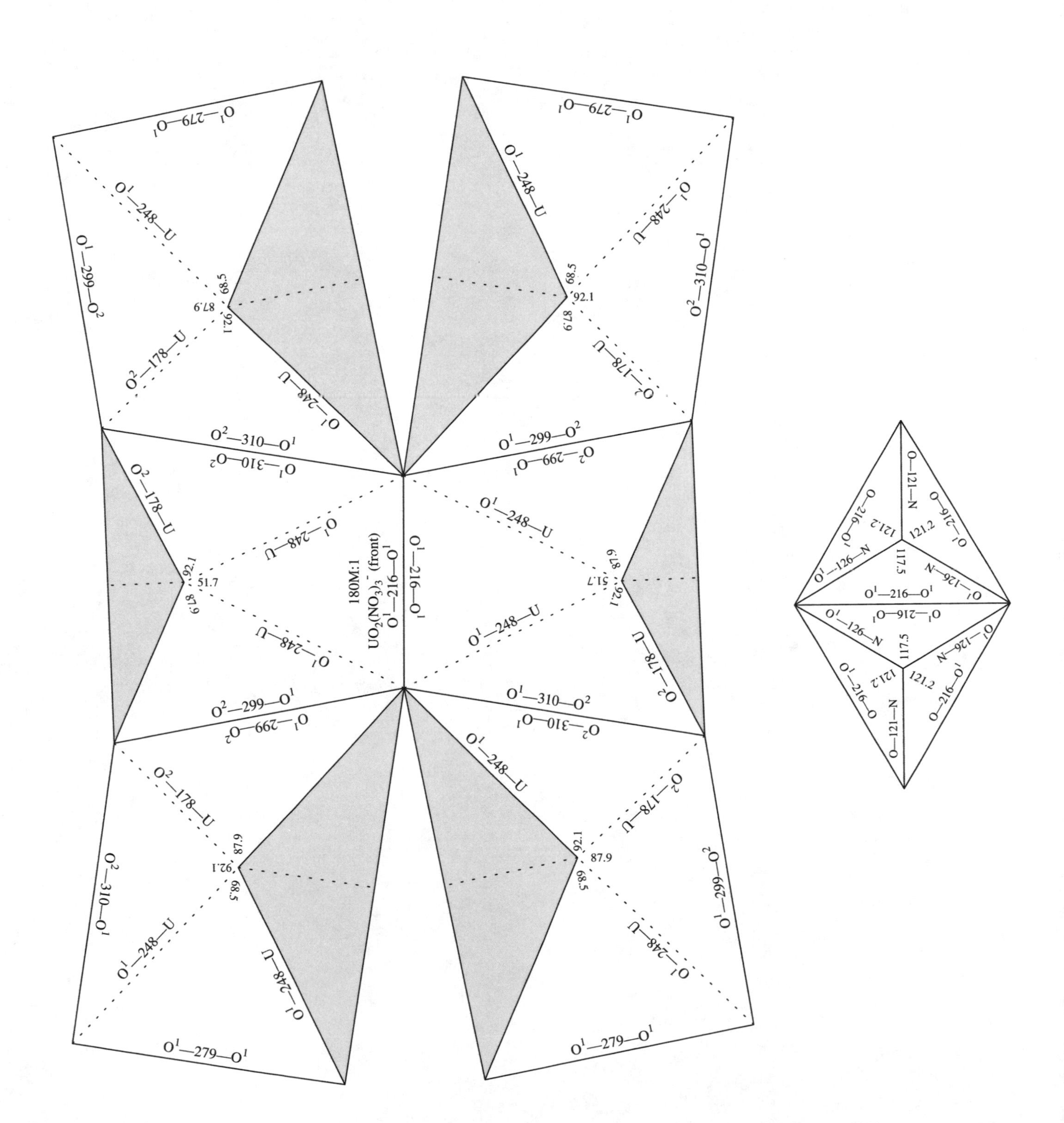

uranyl nitrate ion (back half) $UO_2(NO_3)_3^-$

shape: hexagonal bipyramid units: pm scale: 180,000,000:1

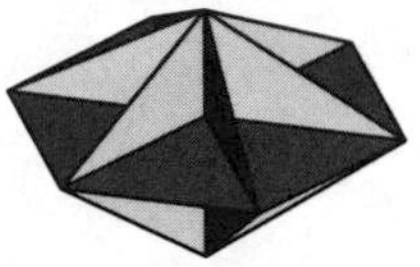

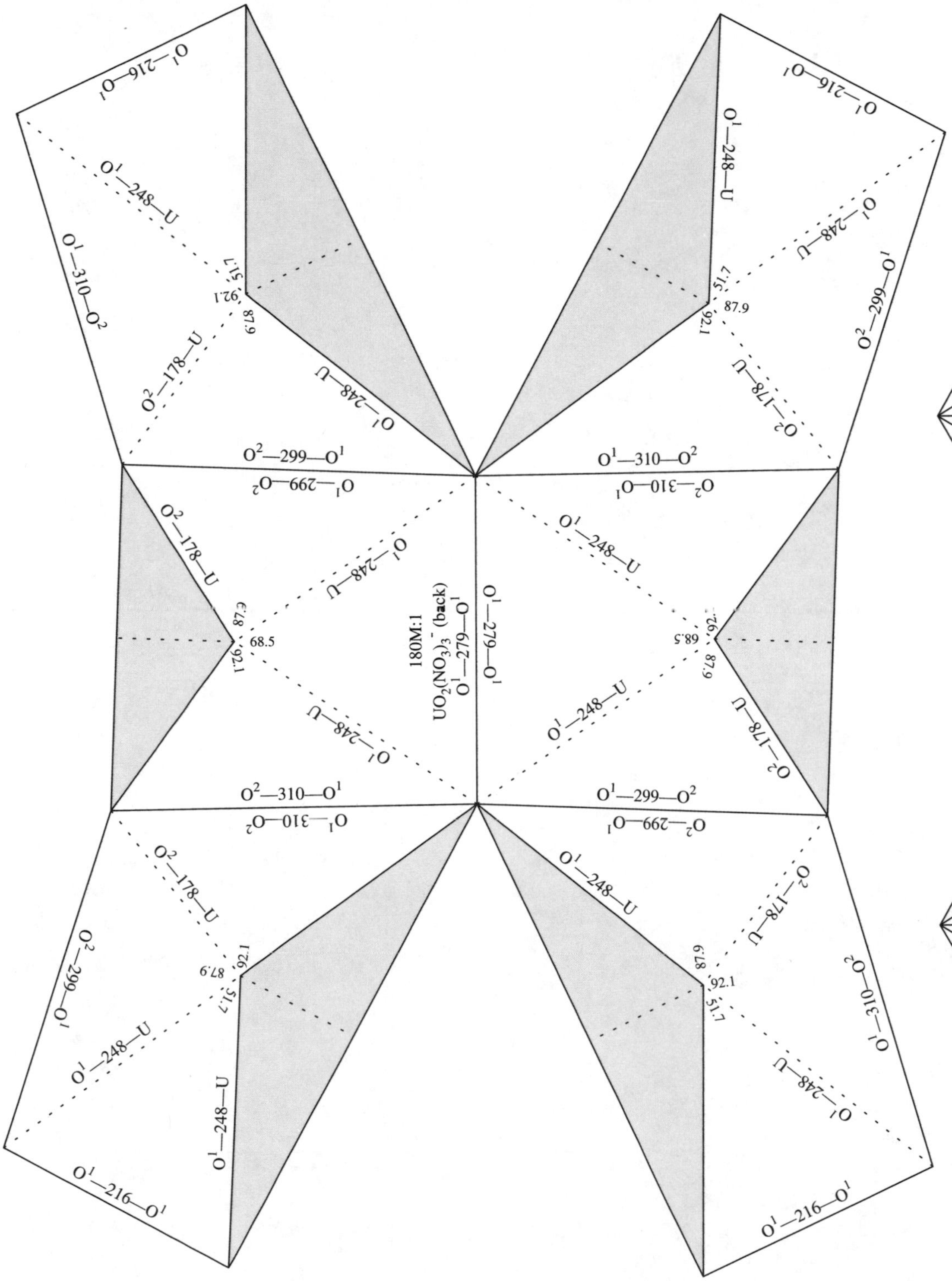

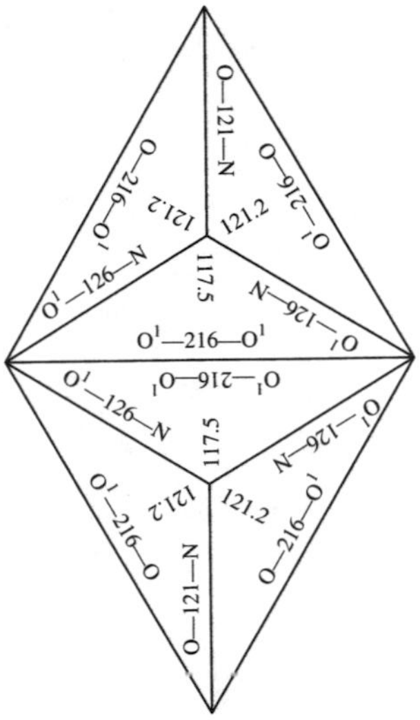

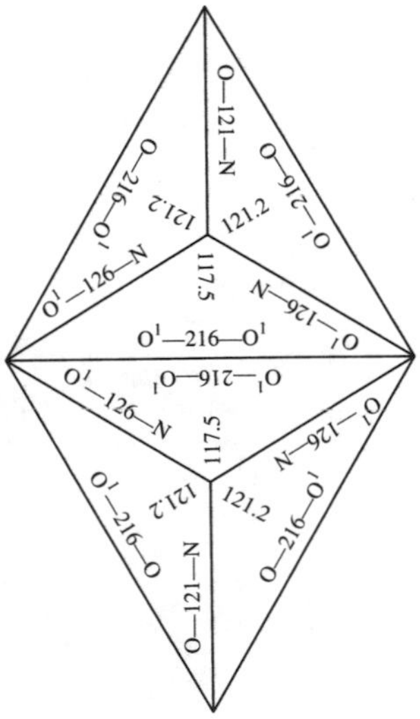

shape: triangulated dodecahedron units: pm scale: 180,000,000:1

planes are perpendicular

Make two, connecting on each model only the F^1–F^2 edges. Then combine the two models along the F^2–F^2 edges.

octafluorozirconate(IV) ion (second copy) ZrF_8^{-4}

shape: triangulated dodecahedron units: pm scale: 180,000,000:1

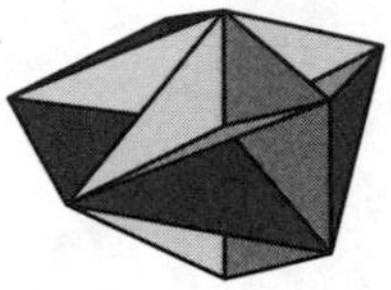

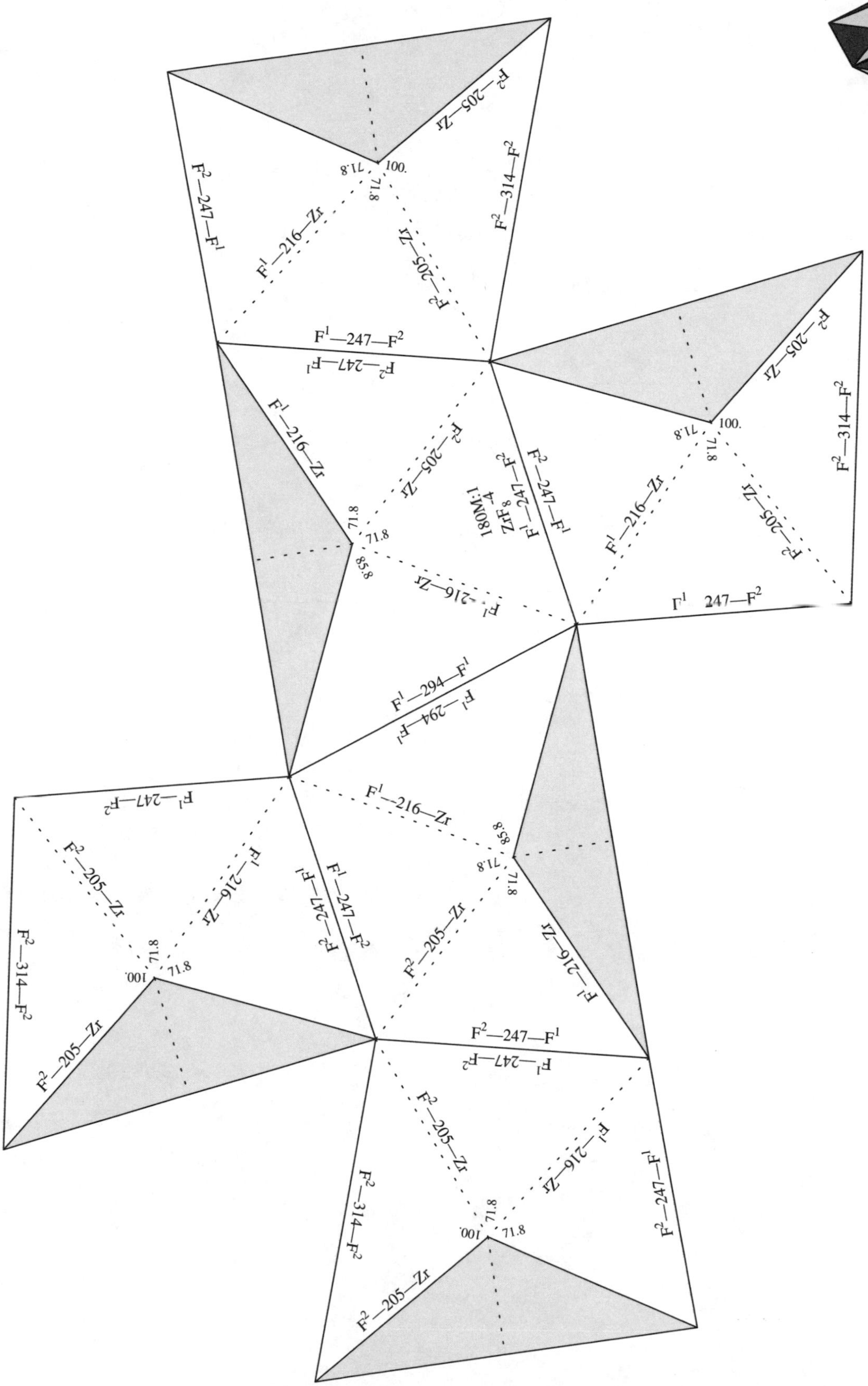

octafluorozirconate(IV) ion (in $[Cu(H_2O)_6]_2ZrF_8$) ZrF_8^{-4}

shape: square antiprism units: pm scale: 180,000,000:1

This is the top part.

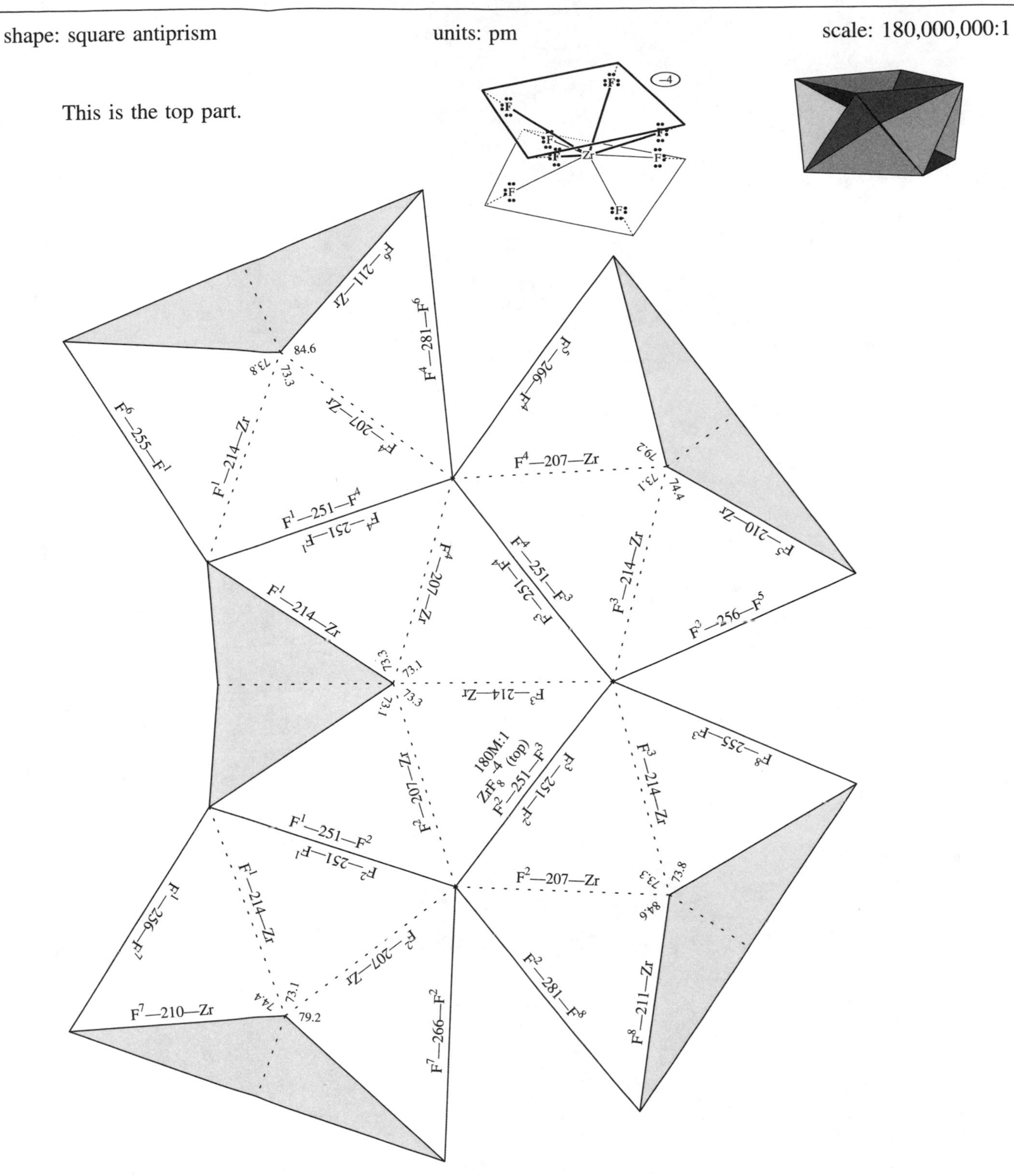

octafluorozirconate(IV) ion (bottom part) ZrF_8^{-4}

shape: square antiprism units: pm scale: 180,000,000:1

Tape this and the top part together by interweaving outer faces.

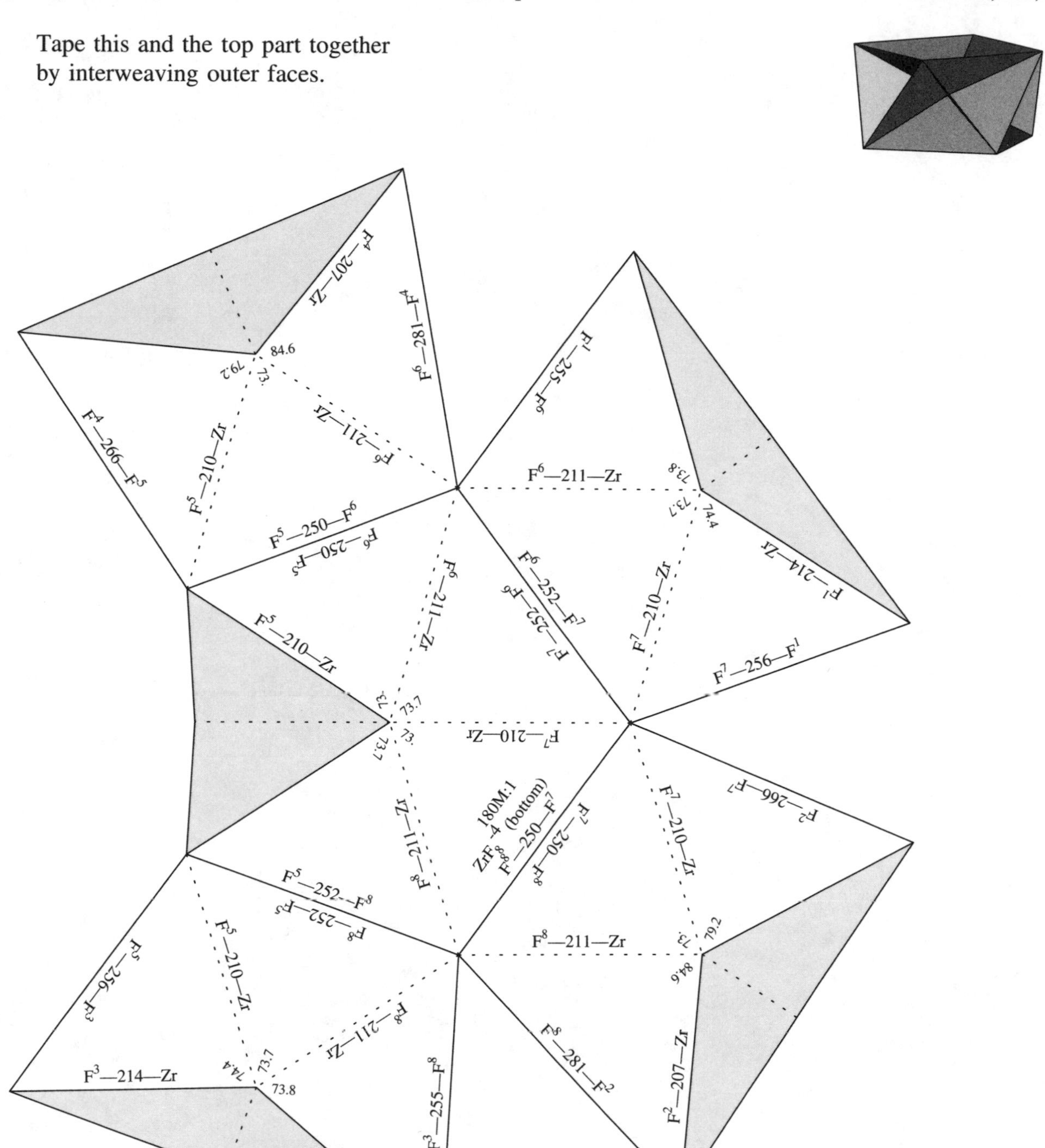

octafluoroxenate(VI) ion (in $(NO)_2XeF_8$) XeF_8^{-2}

shape: square antiprism units: pm scale: 180,000,000:1

This is the top part.

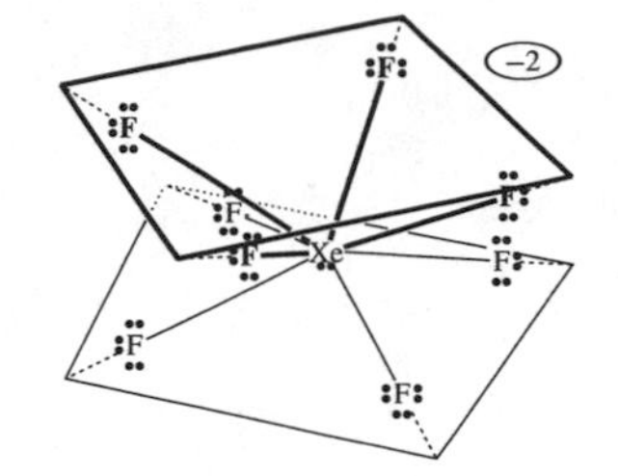

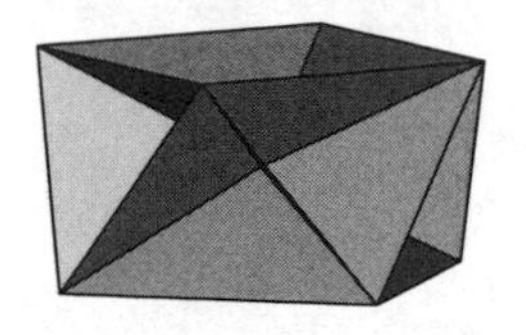

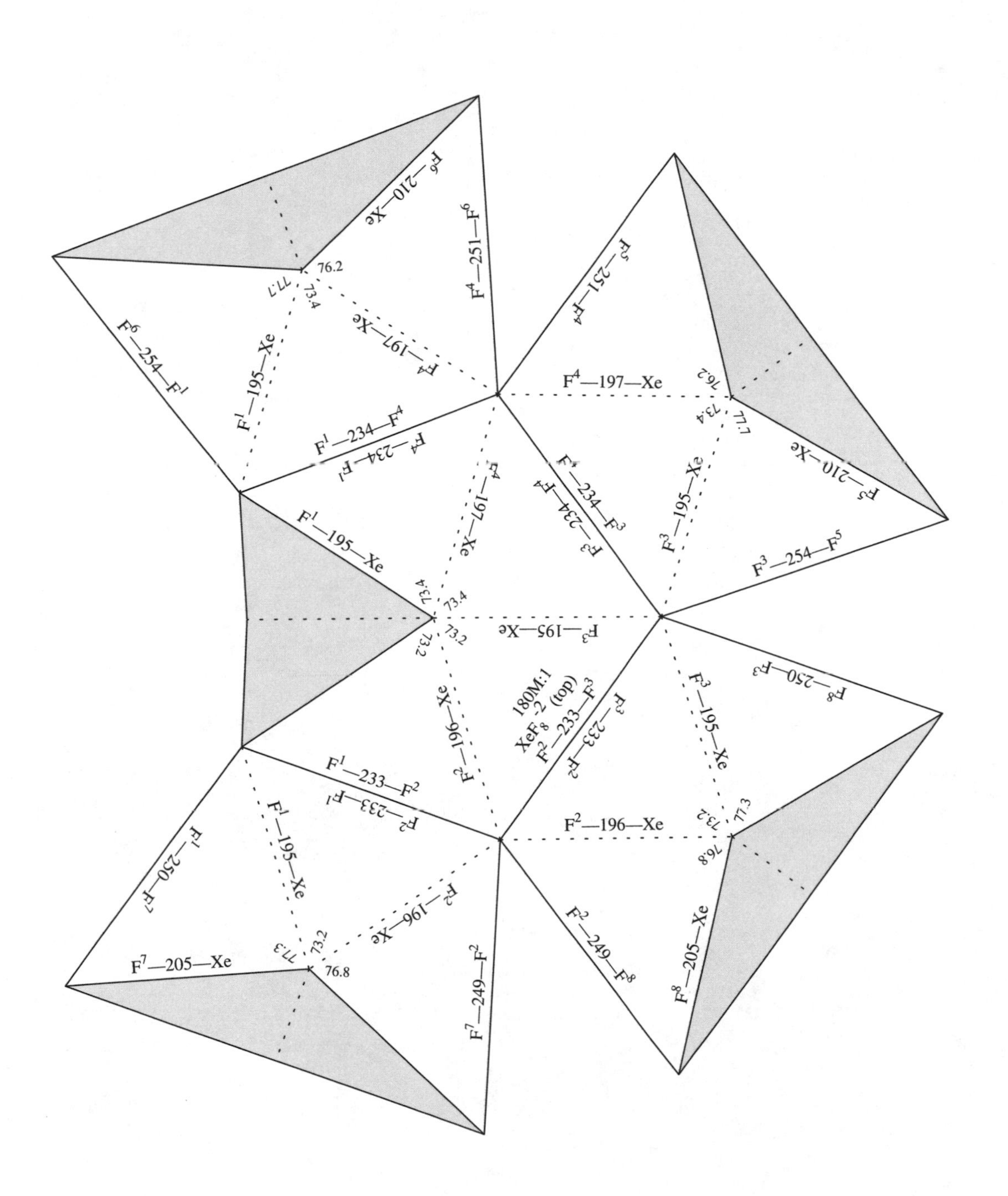

octafluoroxenate(VI) ion (bottom part) XeF_8^{-2}

shape: square antiprism units: pm scale: 180,000,000:1

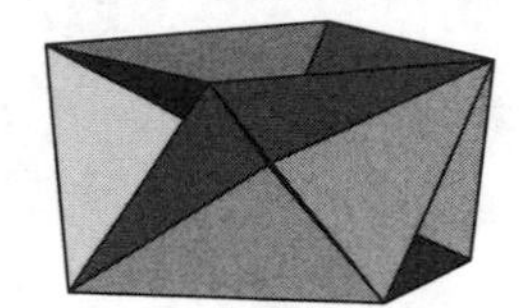

Tape this and the top part together by interweaving outer faces.

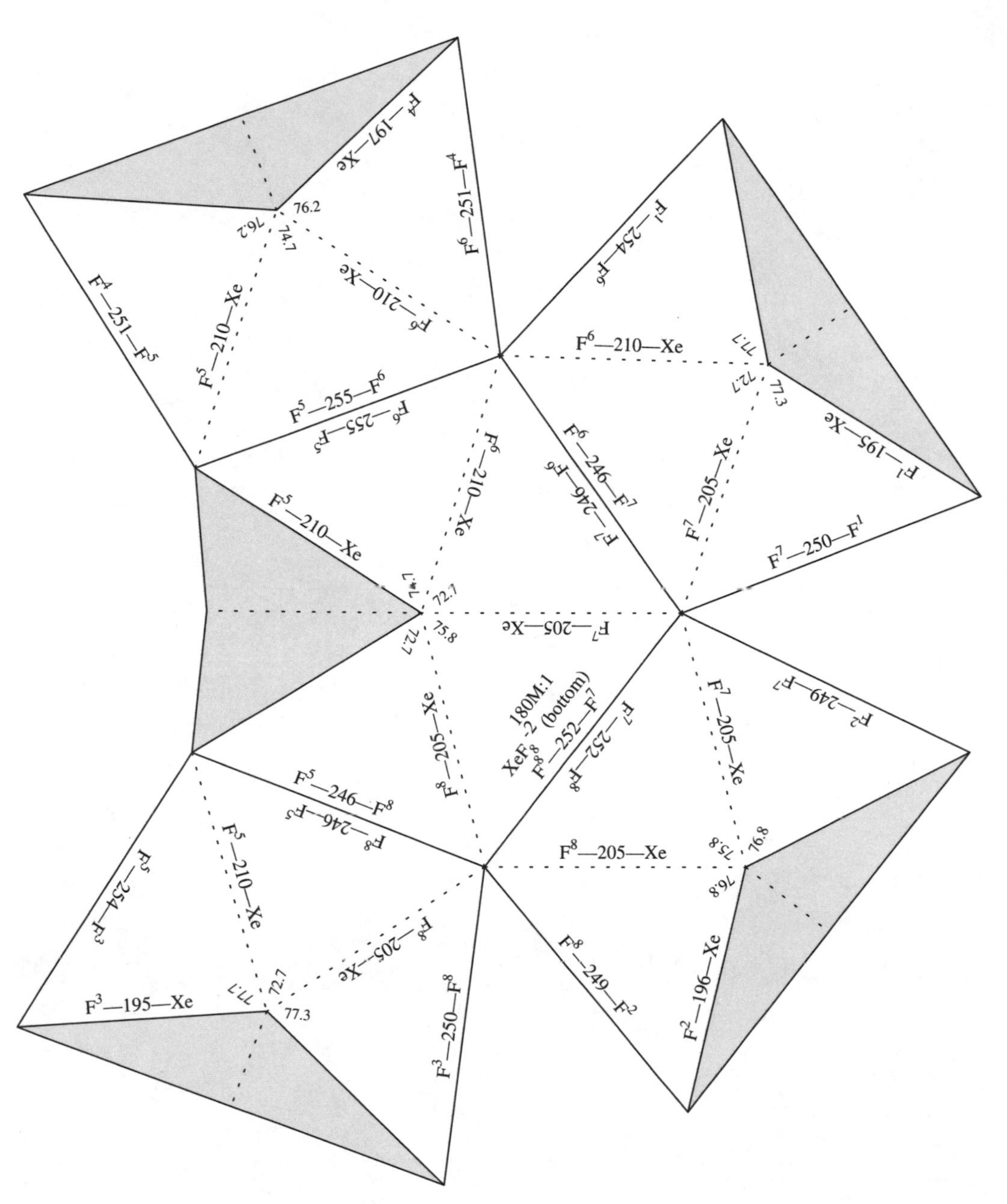

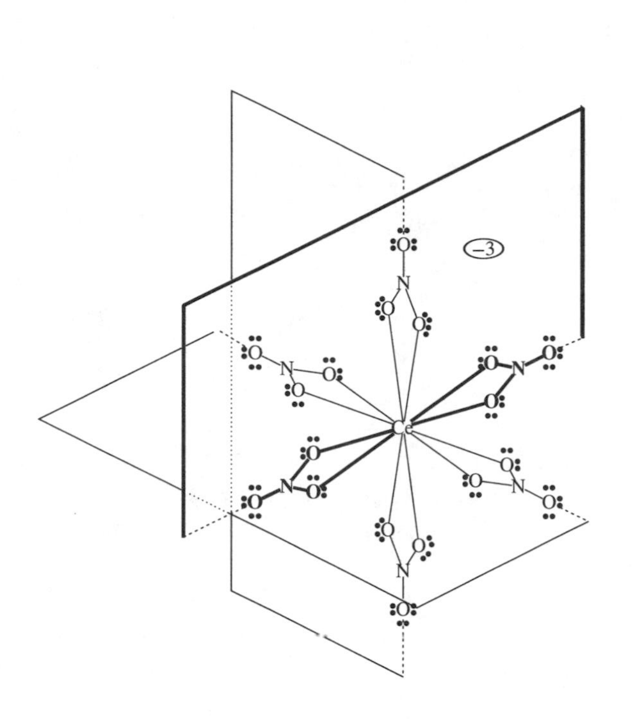

Front view

"Tabs" on these figures are the NO_3^- groups. Notice that the front and back sides of this ion are not quite identical. They are (almost) *mirror images* of each other. Thus, although each half is propeller-shaped, this "boat" would go nowhere!

Back view

hexanitrocerate(III) ion (front and back) $Ce(NO_3)_6^{-3}$

shape: icosahedron units: pm scale: 125,000,000:1

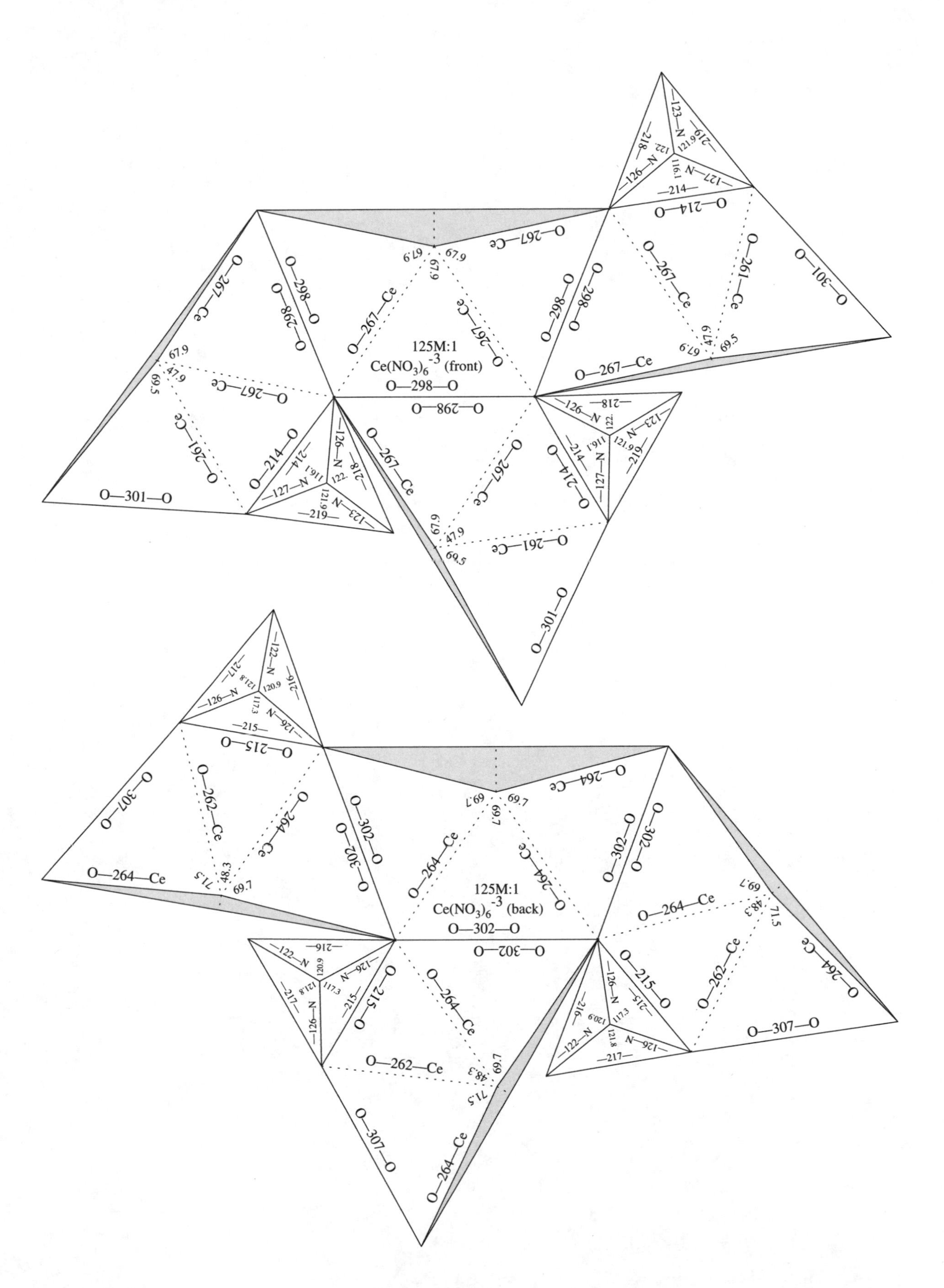

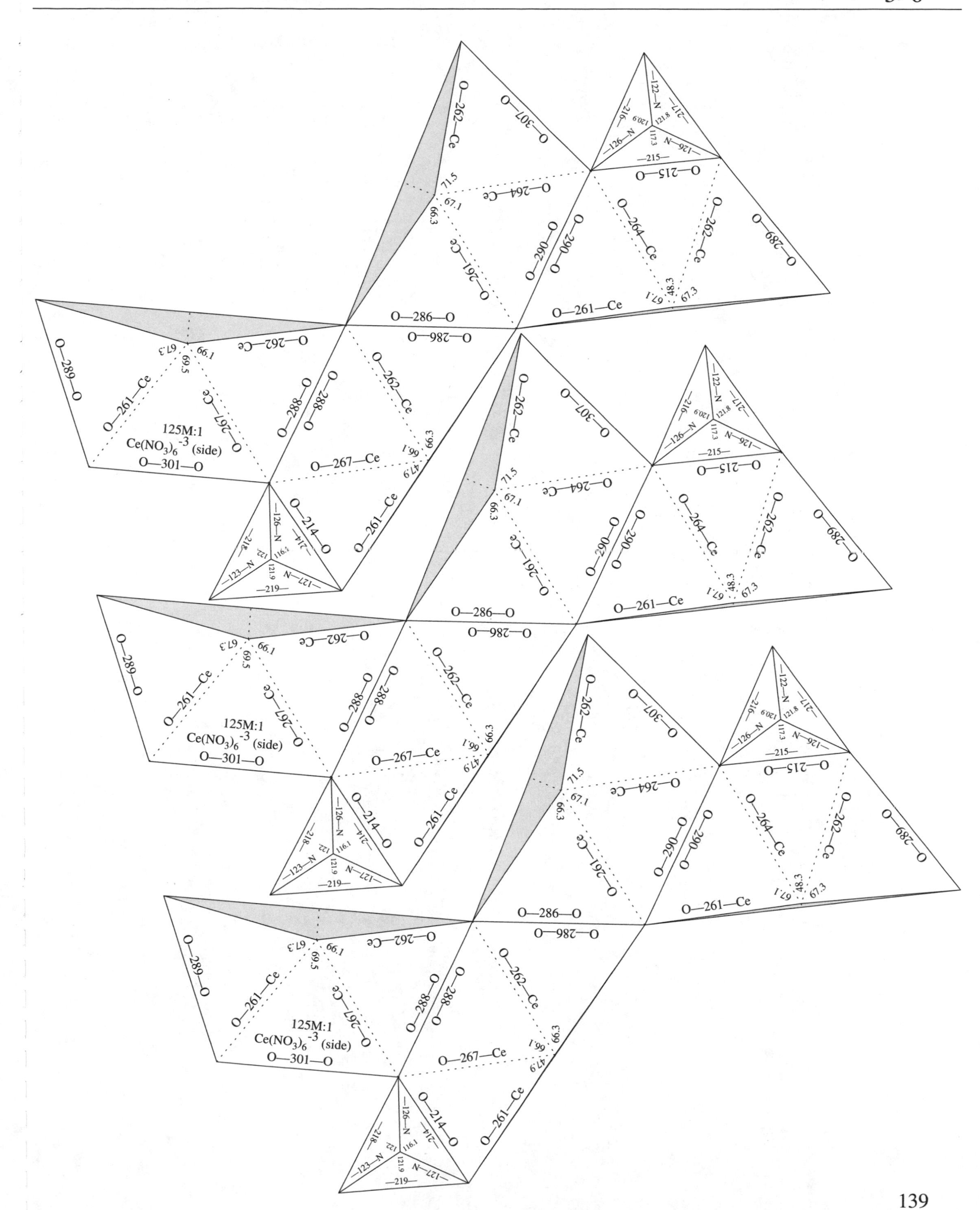
125M:1
Ce(NO3)6-3 (side)
O—301—O
O—286—O
O—267—Ce
O—261—Ce
125M:1
Ce(NO3)6-3 (side)
O—301—O
O—286—O
O—267—Ce
O—261—Ce
125M:1
Ce(NO3)6-3 (side)
O—301—O
O—286—O
O—267—Ce
O—261—Ce

PART 4

More Complex Molecules and Ions

In this part of *Molecular Origami*, the focus is on molecules that arise by combining two or more "central atoms." Of course, there are countless ways this can be done, but five important classes of interaction include:

- **Chaining.** Here the central atom of one unit becomes the surrounding atom of another. Patterns for three such molecules, $BH_3{\cdot}PF_3$ (page 43), CH_3CH_3 (page 49), and CH_3NH_2 (pages 17 and 53), have already been given in Part 1.
- **Edge Sharing.** A common way electron-deficient molecules can fill out their valence is by sharing two atoms with another molecule. B_2H_6 (page 143) is a *dimer* that illustrates the edge-sharing mode of interaction.
- **Face Sharing.** Many inorganic molecules and ions involve octahedra that share faces (three outer atoms). $Fe_2(CO)_9$ (page 145) illustrates what distortions can occur when two octahedra share faces.
- **Clustering.** This is really just a fancy name for chaining primarily involving inorganic systems. Clusters are molecules that consist of a group of at least three directly connected main atoms with no specific one designated "central." The patterns here focus on the main atoms in the cluster and include P_4 (page 149), $B_{12}H_{12}^{-2}$ (page 151), and C_{60} (page 155).
- **Corner Sharing.** In this mode, one outer atom links two central atoms. A beautiful example of corner sharing is quartz, which forms an extended network. Quartz is included in Part 5.

Refer to the introduction for hints about putting these units together.

diborane B_2H_6

shape: edge-linked tetrahedra

units: pm

scale: 300,000,000:1

Connect these two halves along the H^2–H^2 edge.

diiron nonacarbonyl $Fe_2(CO)_9$

shape: face-sharing octahedra units: pm scale: 200,000,000:1

Two copies are required.

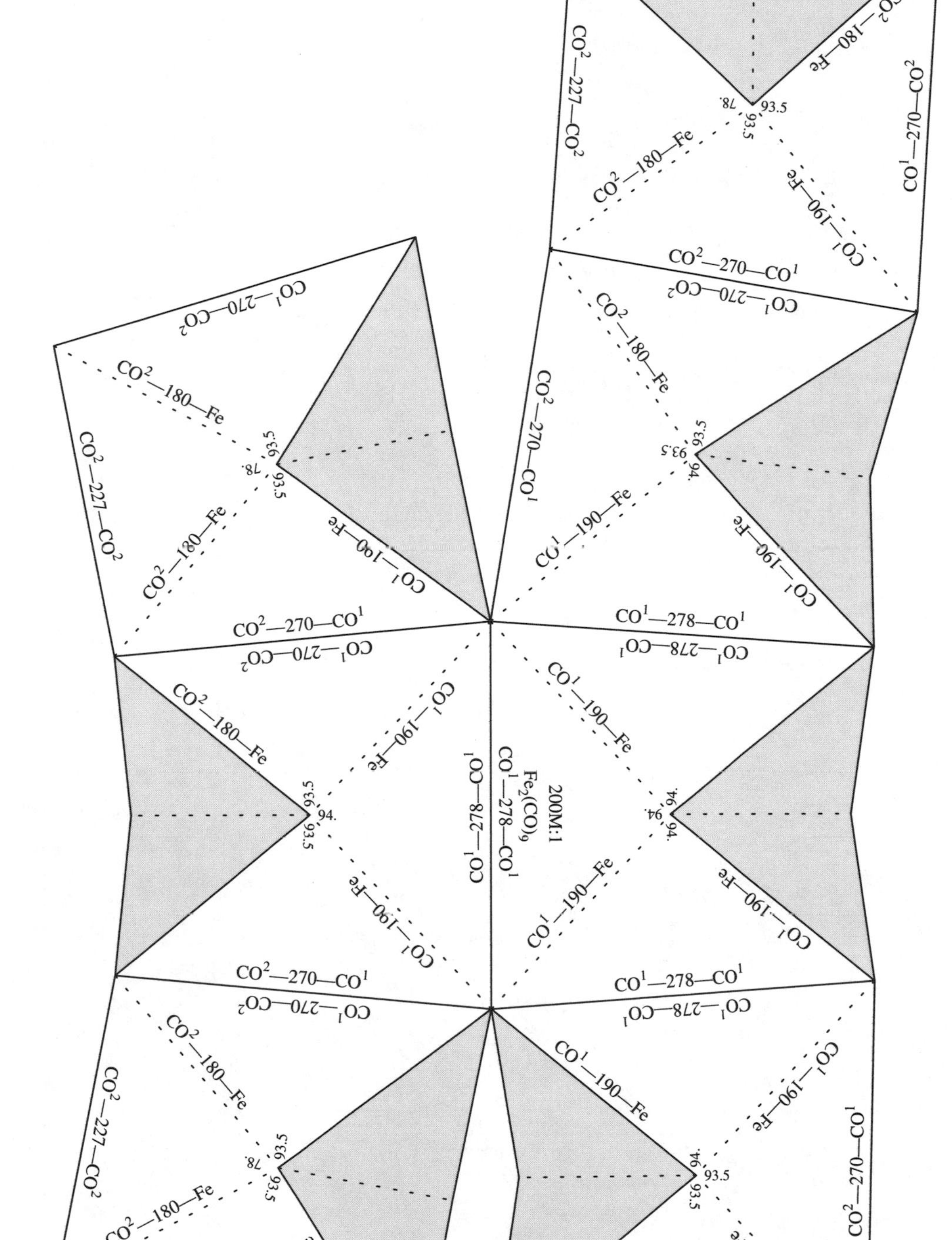

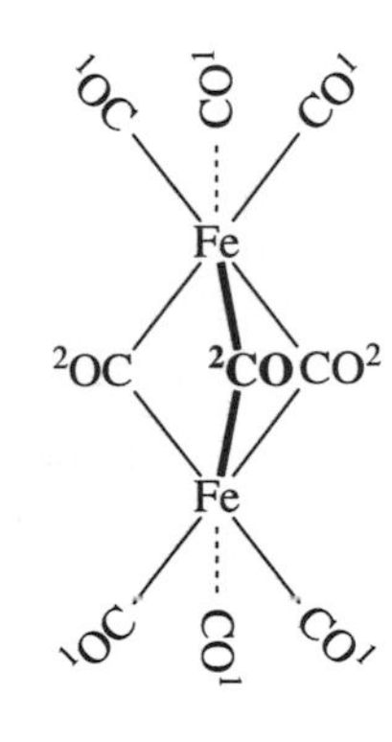

diiron nonacarbonyl (second copy) $Fe_2(CO)_9$

shape: face-sharing octahedra | units: pm | scale: 200,000,000:1

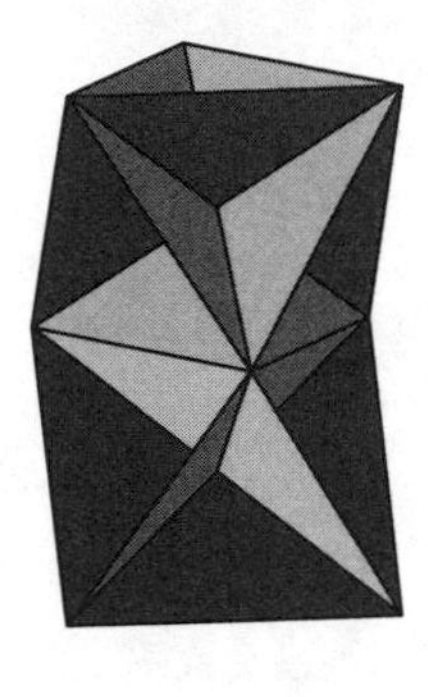

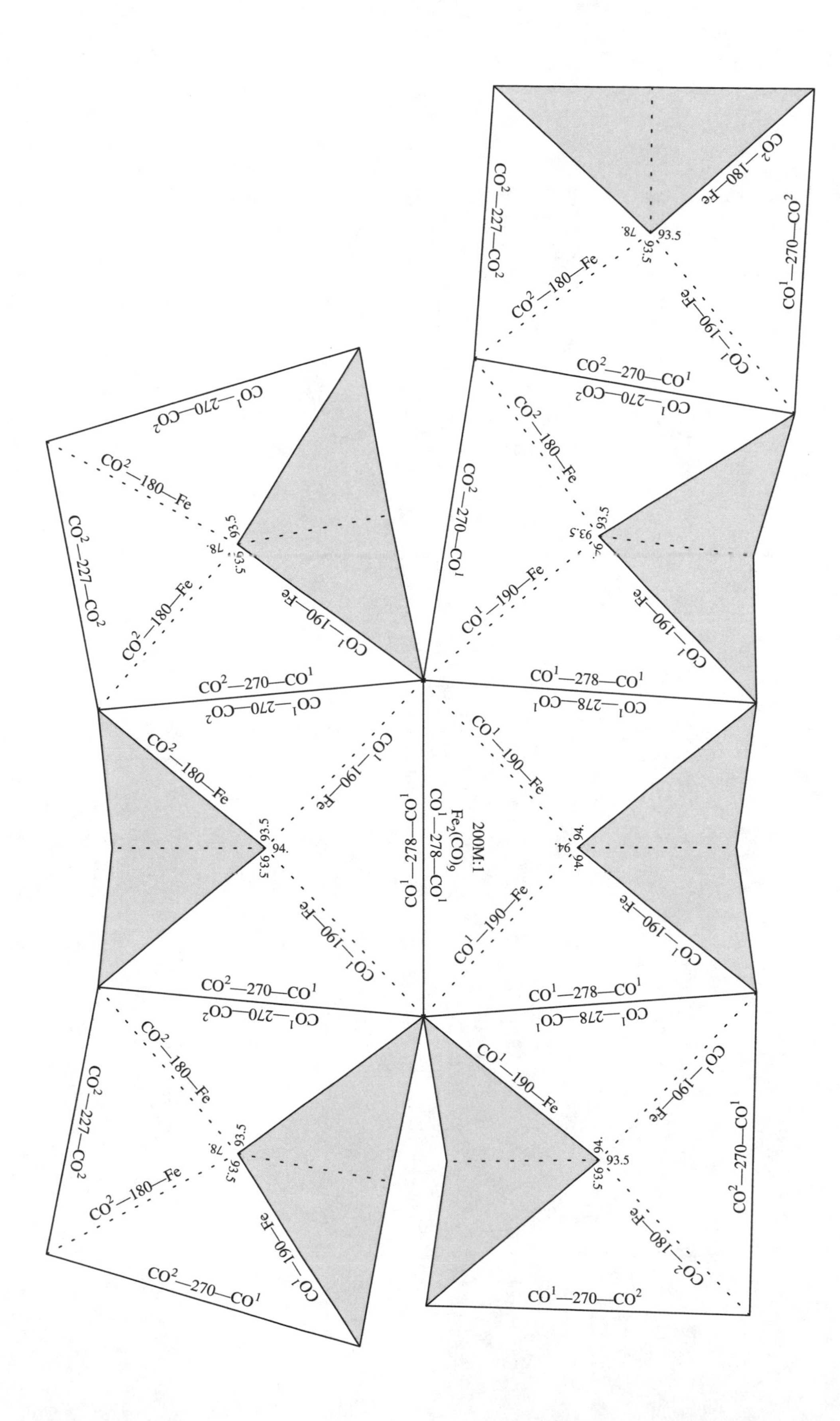

tetraphosphorus P_4

shape: tetrahedral cluster | units: pm | scale: 300,000,000:1

The lower set of shaded triangles forms two tabs that slip into pockets formed by the upper set of shaded triangles.

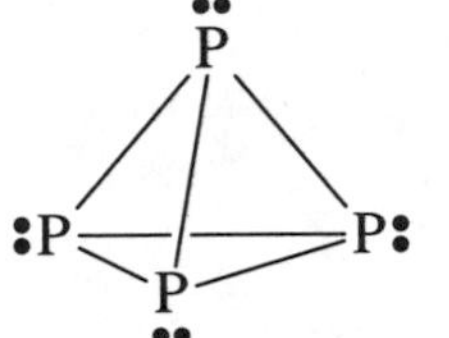
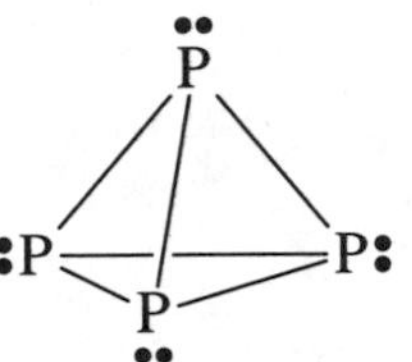

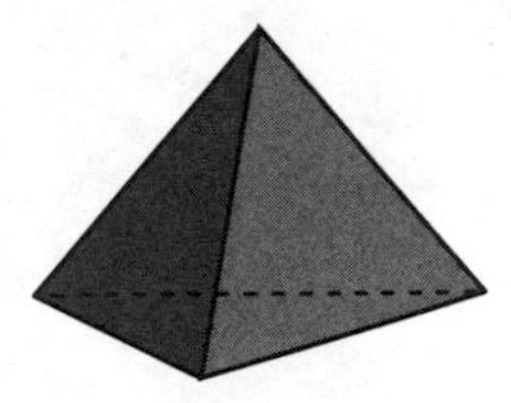

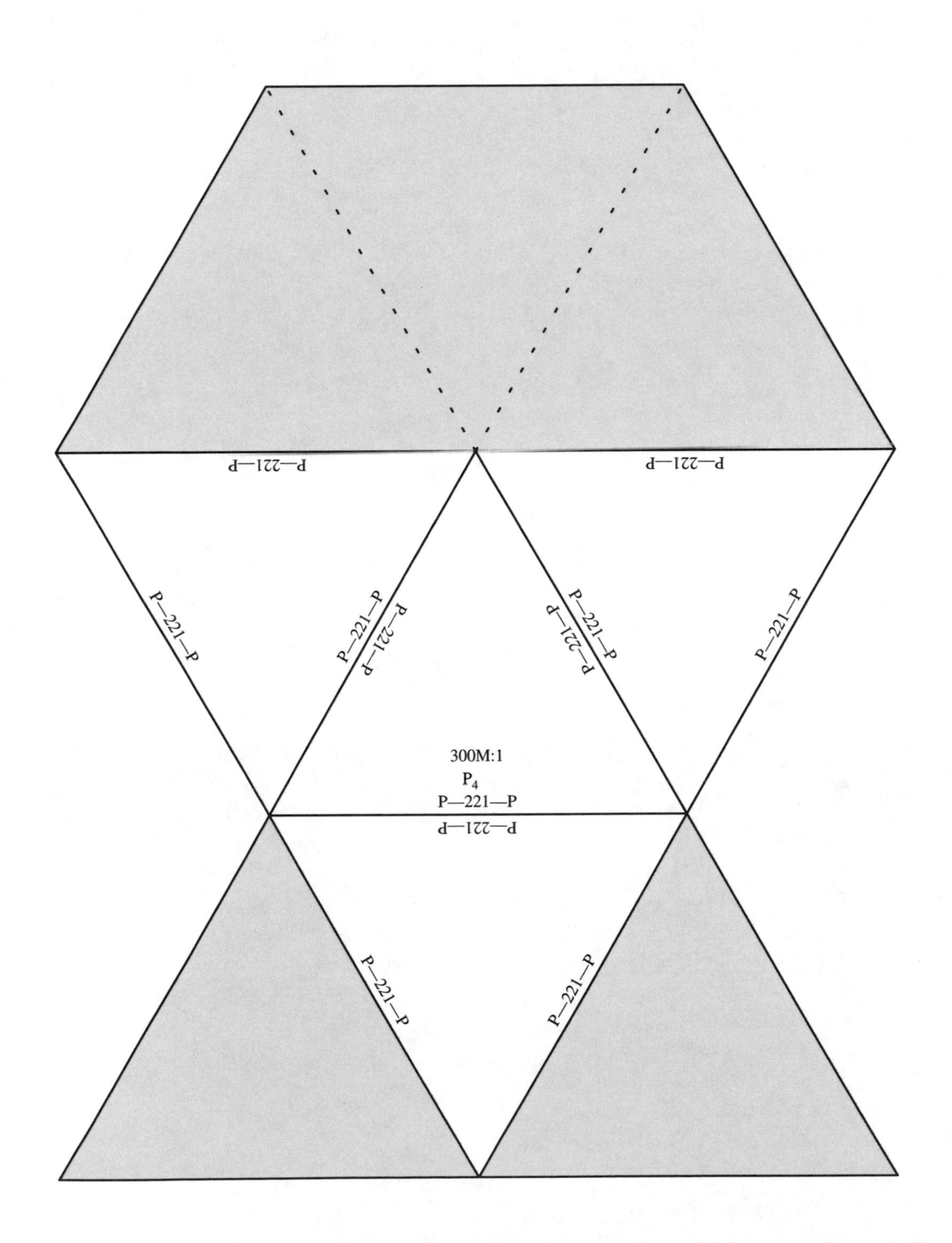

dodecaborane ion $B_{12}H_{12}^{-2}$

shape: icosahedron units: pm scale: 300,000,000:1

The B–B distances in this model are average distances; some are actually 178 pm, and some are 175 pm. The B–H distances were not determined. The "central atom" is actually nonexistent—adding it simply makes the model stronger and look better. (Its official mathematical name is the *great dodecahedron*.)

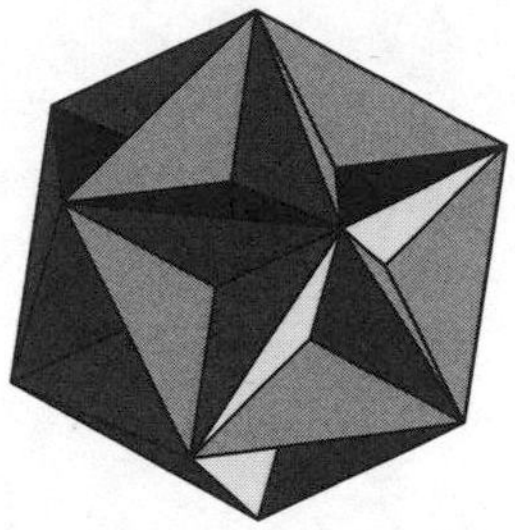

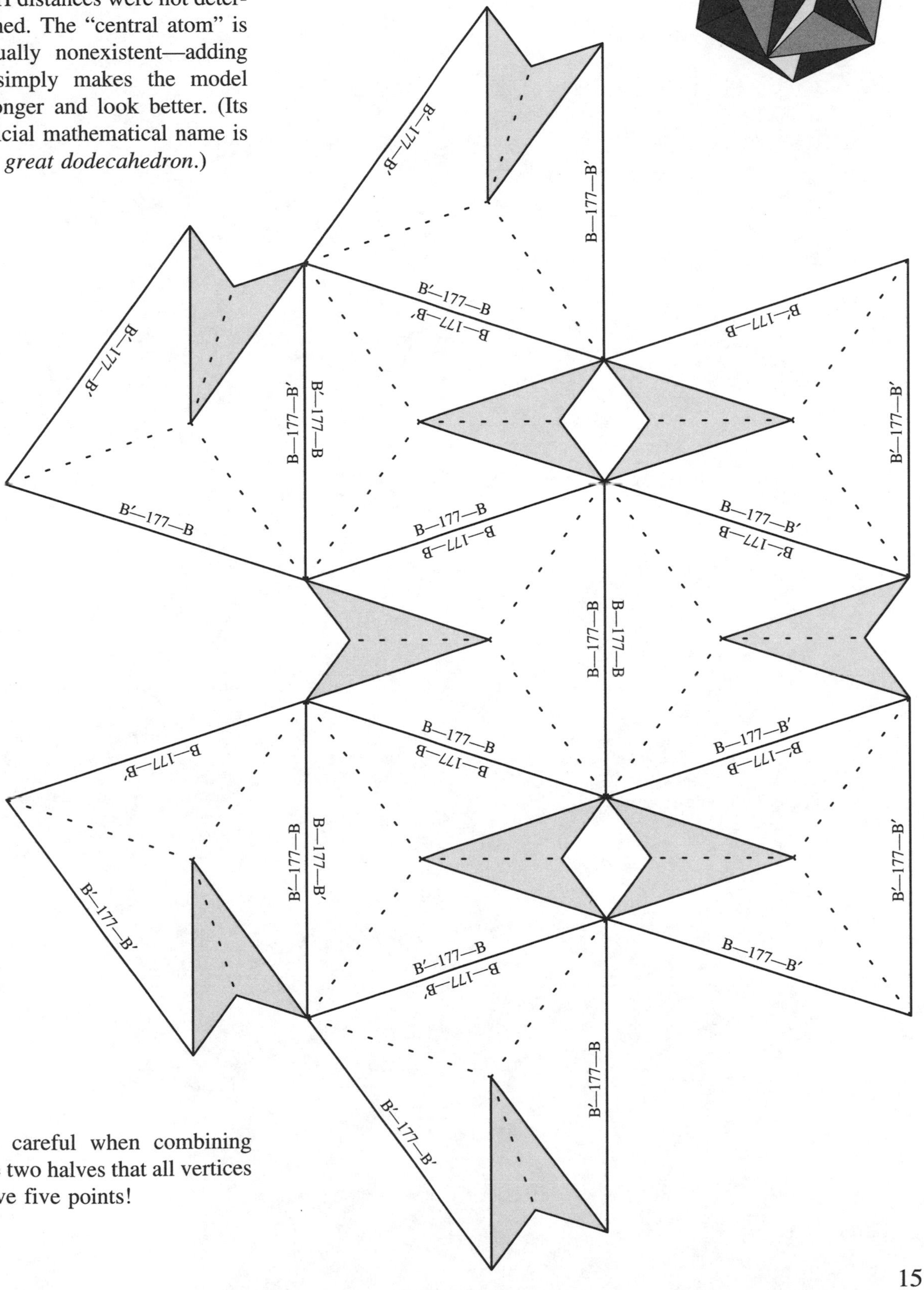

Be careful when combining the two halves that all vertices have five points!

dodecaborane ion (second copy) $B_{12}H_{12}^{-2}$

shape: great dodecahedron units: pm scale: 300,000,000:1

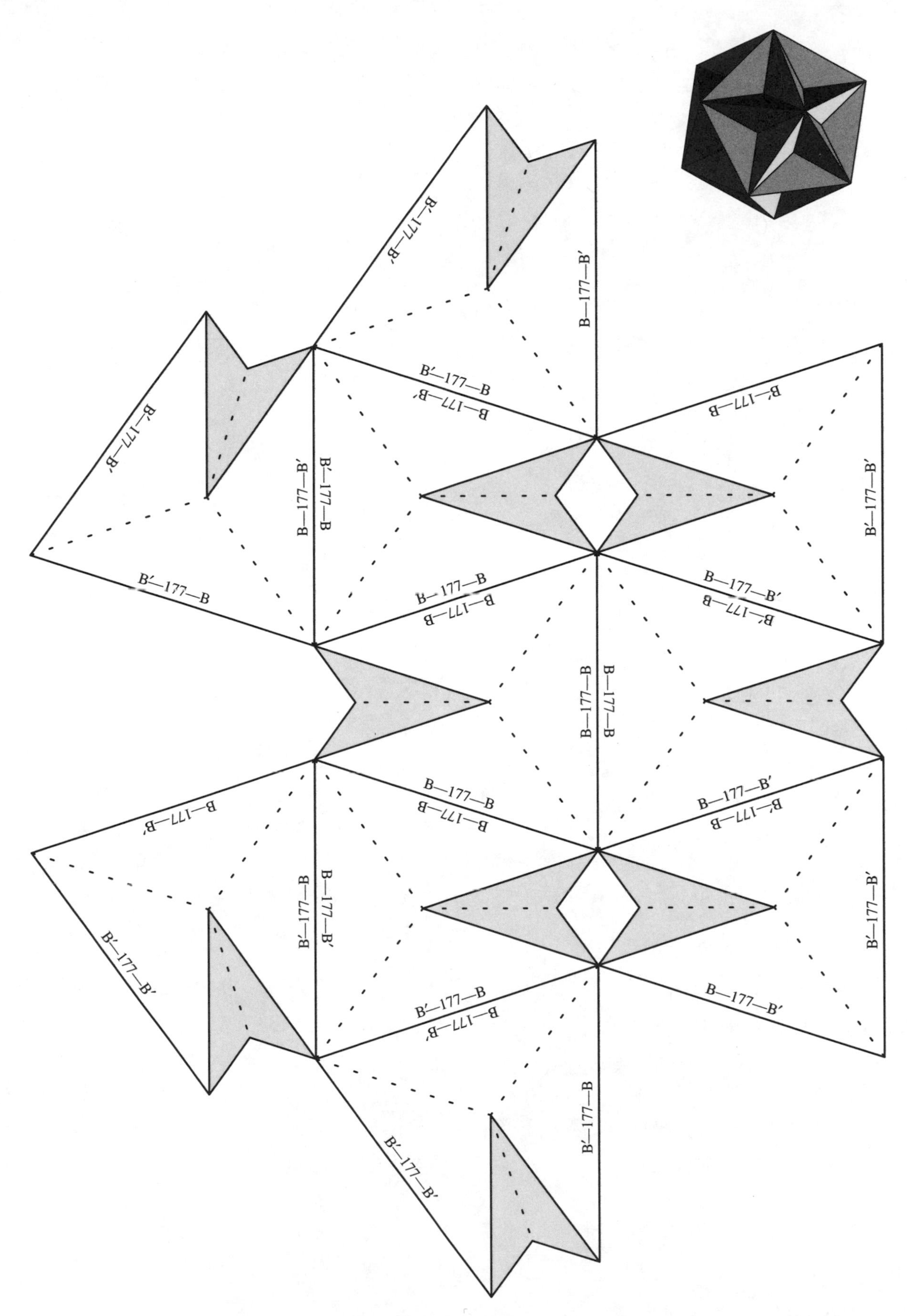

buckminsterfullerene C_{60}

shape: truncated icosahedron units: pm scale: 150,000,000:1

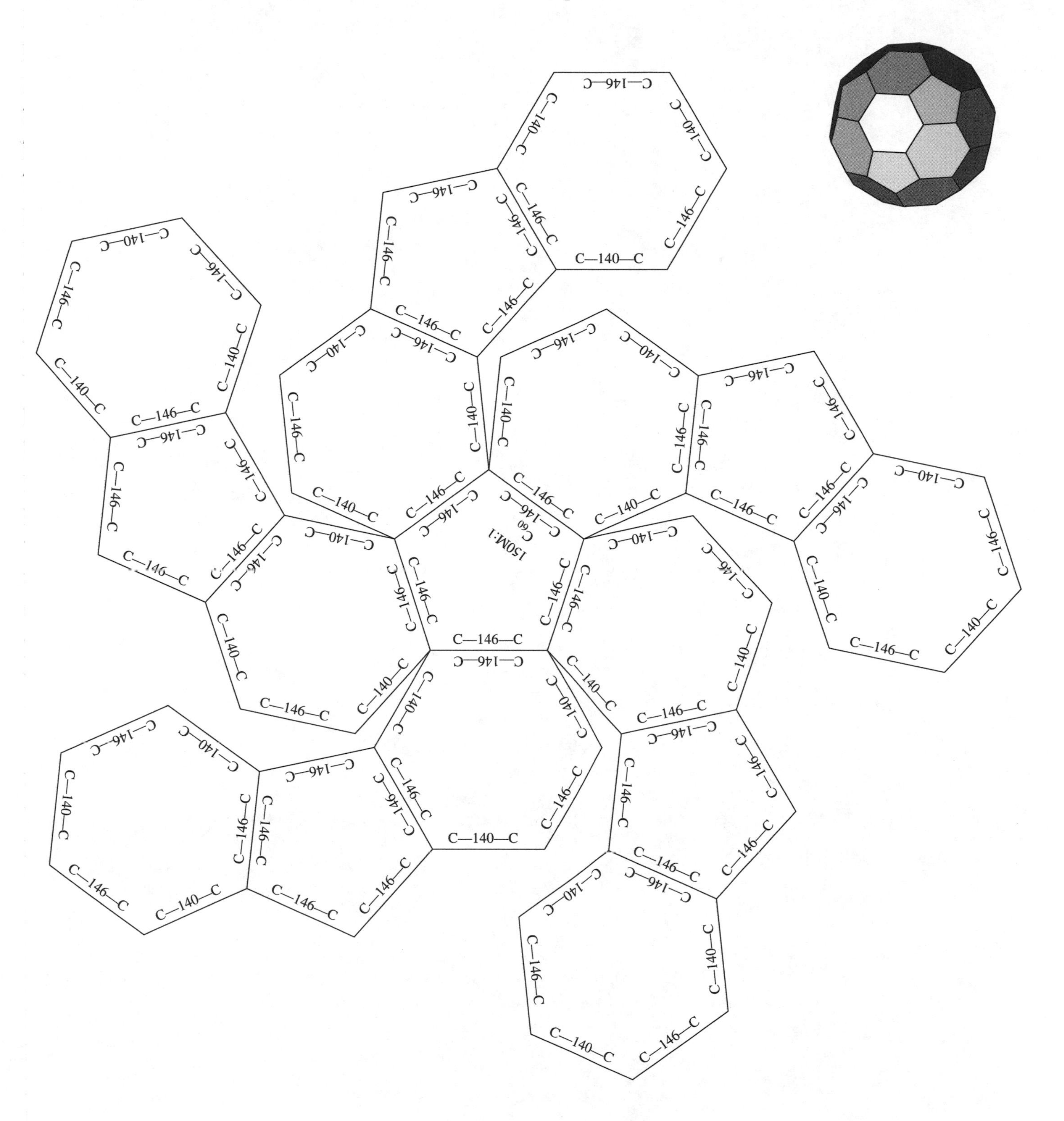

Note the two *different* C–C distances. Most aromatic compounds have C–C distances around 139 pm.

buckminsterfullerene (second copy) C_{60}

shape: truncated icosahedron   units: pm   scale: 150,000,000:1

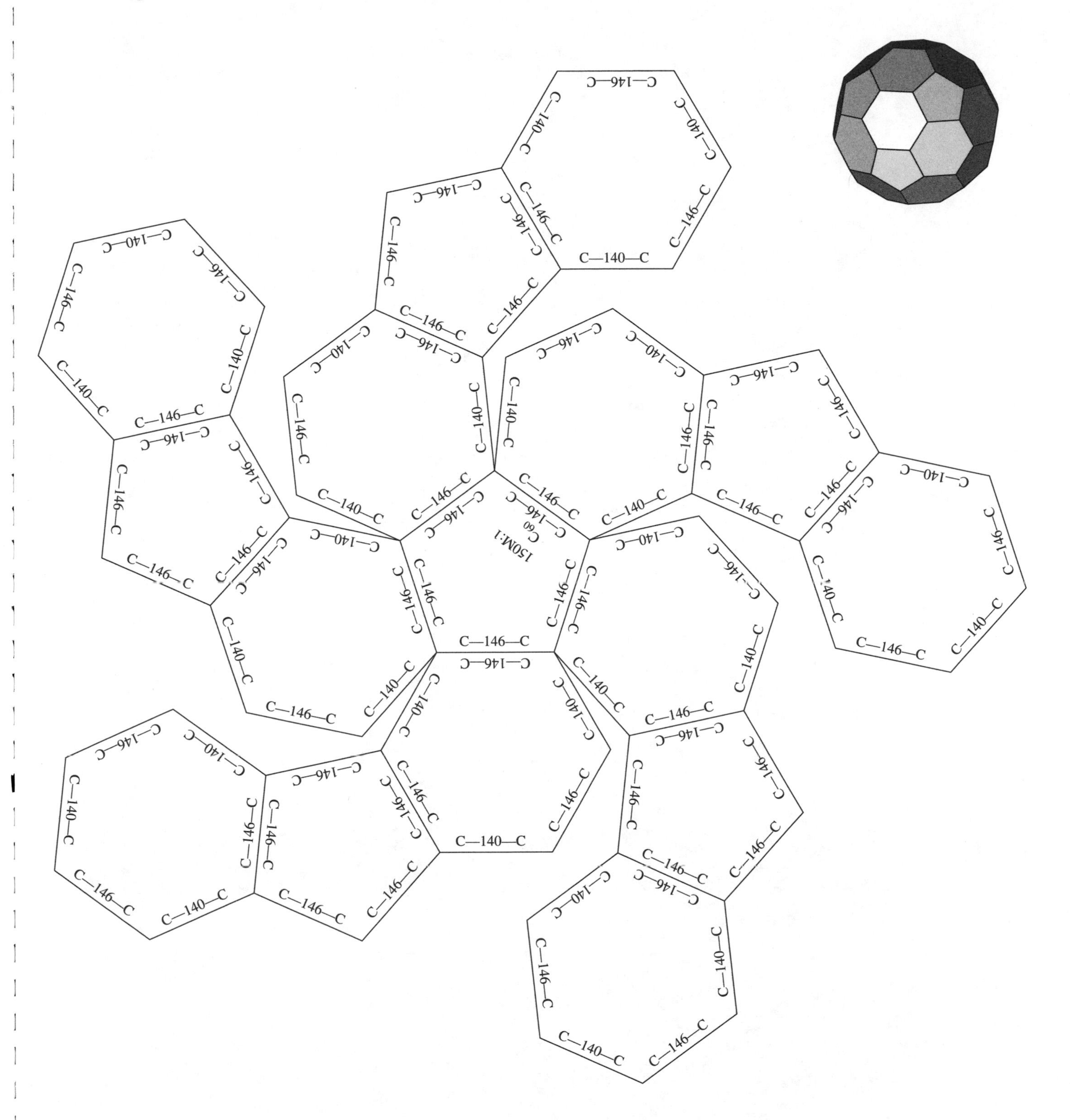

PART 5

Network Solids

Network solids are solids that cannot be dissected into individual unit molecules without breaking covalent bonds. Thus, like their cousins the *ionic* solids, network solids consist of one giant molecule that goes on in at least two directions. (Also related to network solids are *polymers*, which generally have more of a one-dimensional character.) The example given here, quartz, is composed of silicon atoms linked by tetrahedrally arranged oxygen atoms:

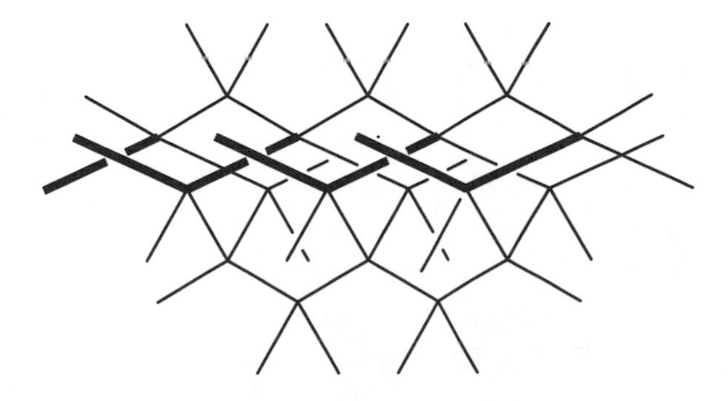

The quartz structure is not easy to depict in two dimensions! In this figure, the intersections of lines are the silicon atoms. One must imagine an oxygen atom in between each pair of silicon atoms. (The detailed structure is much easier to see using a model.)

On this model you will find shaded triangular "tabs" that help you connect one to another. In order to make the model as shown, follow the directions carefully, attaching tabs to each other or to edges of units with care. In most cases, there are extra tabs. You can either just cut them off or use them to continue adding units to the model in different directions.

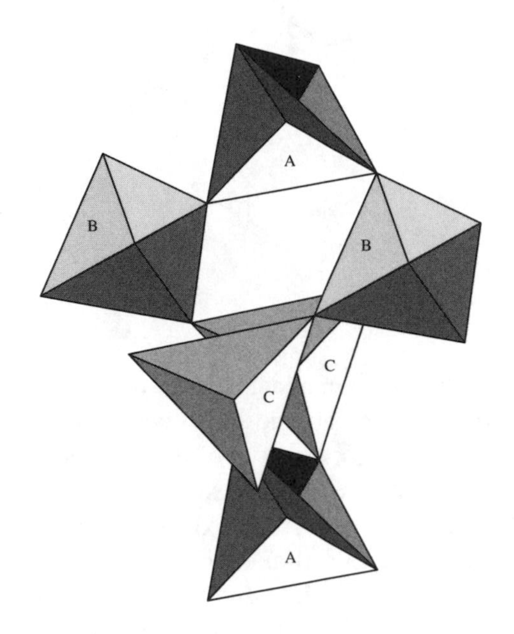

Quartz consists of SiO_4 tetrahedra linked one to another only by their corners. This model consists of six units. Each unit is identical, differing only in orientation. Try to imagine the model extending in all directions by replicating units that you see. (This relationship is very easy to see on the completed model.) Note that the structure of quartz involves a helical twist, which runs upward, through the two A units. Each unit is rotated 120° relative to the one under it. In nature half of the crystals of quartz turn one way and half turn the other way, based on chance. To make the opposite twist, simply assemble the model *inside out*.

The corner-sharing relationship poses a special challenge for model building. However, there is no real problem if you carefully follow the instructions given here. When making the model, attach the units one to another by overlapping triangular tabs with both shaded sides showing. Tabs are numbered 1–6. Each has two sides, and as long as you always put together two tabs with the same number from *different* units (for example, A1 with B1 or B2 with C2, but not A1 with A1 or B2 with B2) you cannot go wrong. The key is to focus on the *tabs*. Their sequence determines the section of the infinite network that you make. Given here are instructions for making the above models three different but equivalent ways and for extending the model in all directions.

Assembling the quartz model

Three distinct ways of producing the model as shown are depicted below.

A-axis unit B-axis unit C-axis unit

For the A-axis unit, for example, attach tabs in the order shown. That is, overlay tab A1 with B1, then B2 with C2, etc., around to B6 with A6. Each unit will have two extra unused tabs. Using these extra tabs, any two of the three loops can be connected to produce the third. The whole structure can be extended this way in any direction. Shown below is a structure built by connecting a C-axis unit with an A-axis unit via tabs 1 and 5, and then connecting *that* unit to a B-axis unit via tabs 3 and 5.

This whole substructure will attach to another copy of itself using tabs 1 and 4 (or 2 and 4 or 2 and 6 or . . . you get the idea!). In effect, you can continue along any path by connecting tabs as long as you take care (a) to not connect like units such as A1 with A1 or B2 with B2 and (b) to watch out for the path reattaching somewhere else on the model instead of adding another subunit.

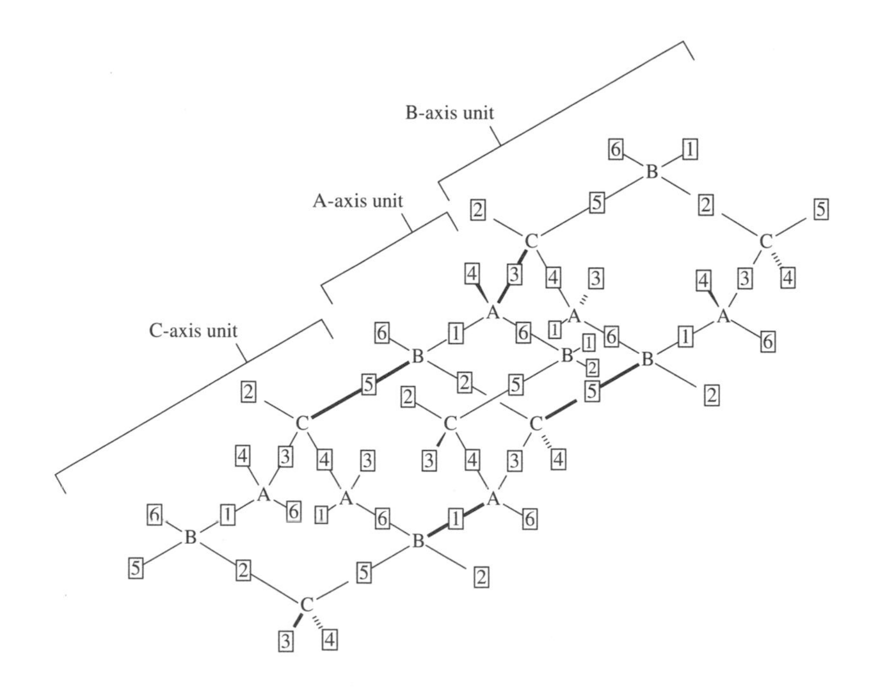

an extensive quartz structure

shape: corner-sharing tetrahedra units: pm scale: 240,000,000:1

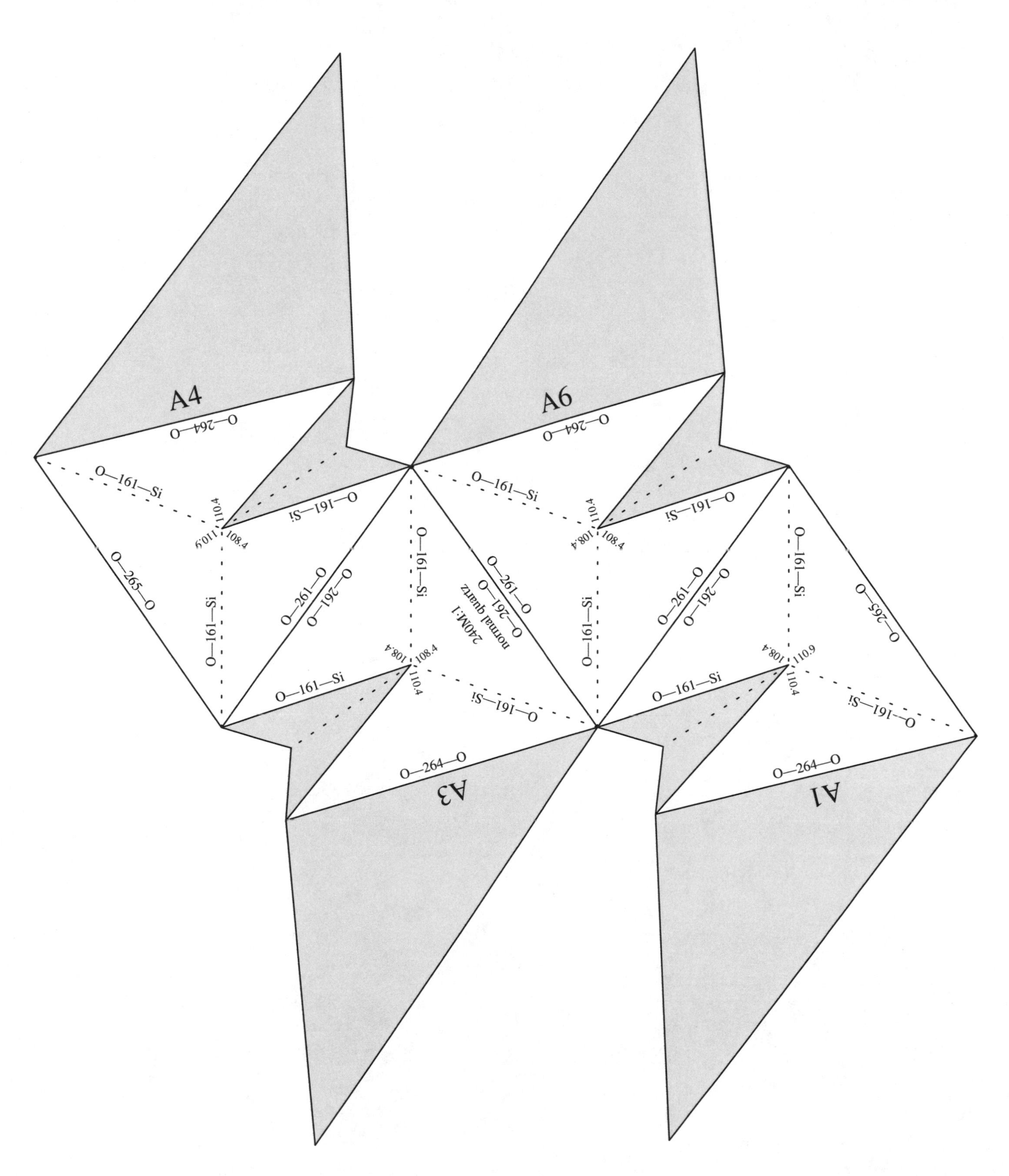

normal (α) quartz (unit A) $(SiO_2)_x$

shape: corner-sharing tetrahedra units: pm scale: 240,000,000:1

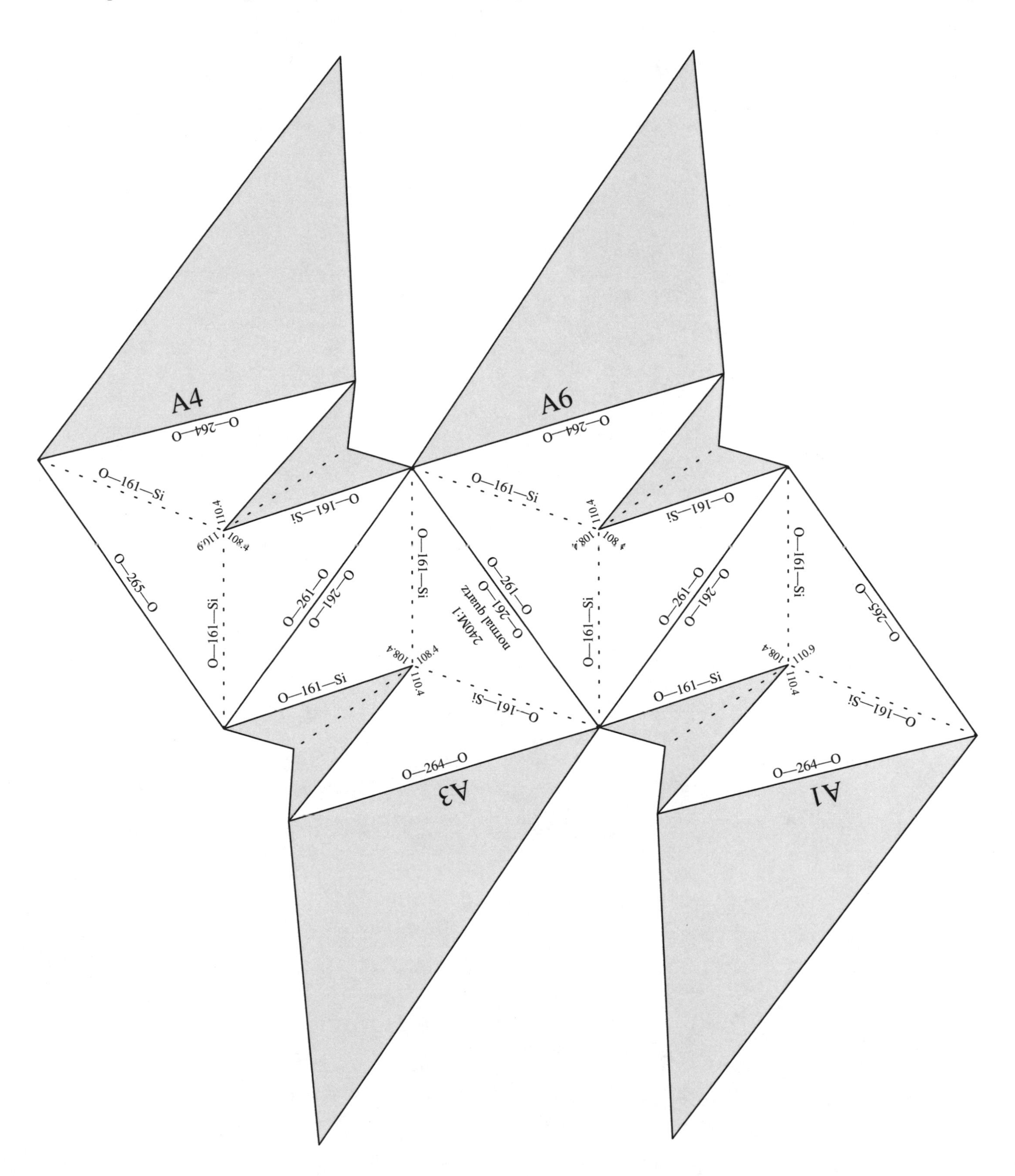

shape: corner-sharing tetrahedra units: pm scale: 240,000,000:1

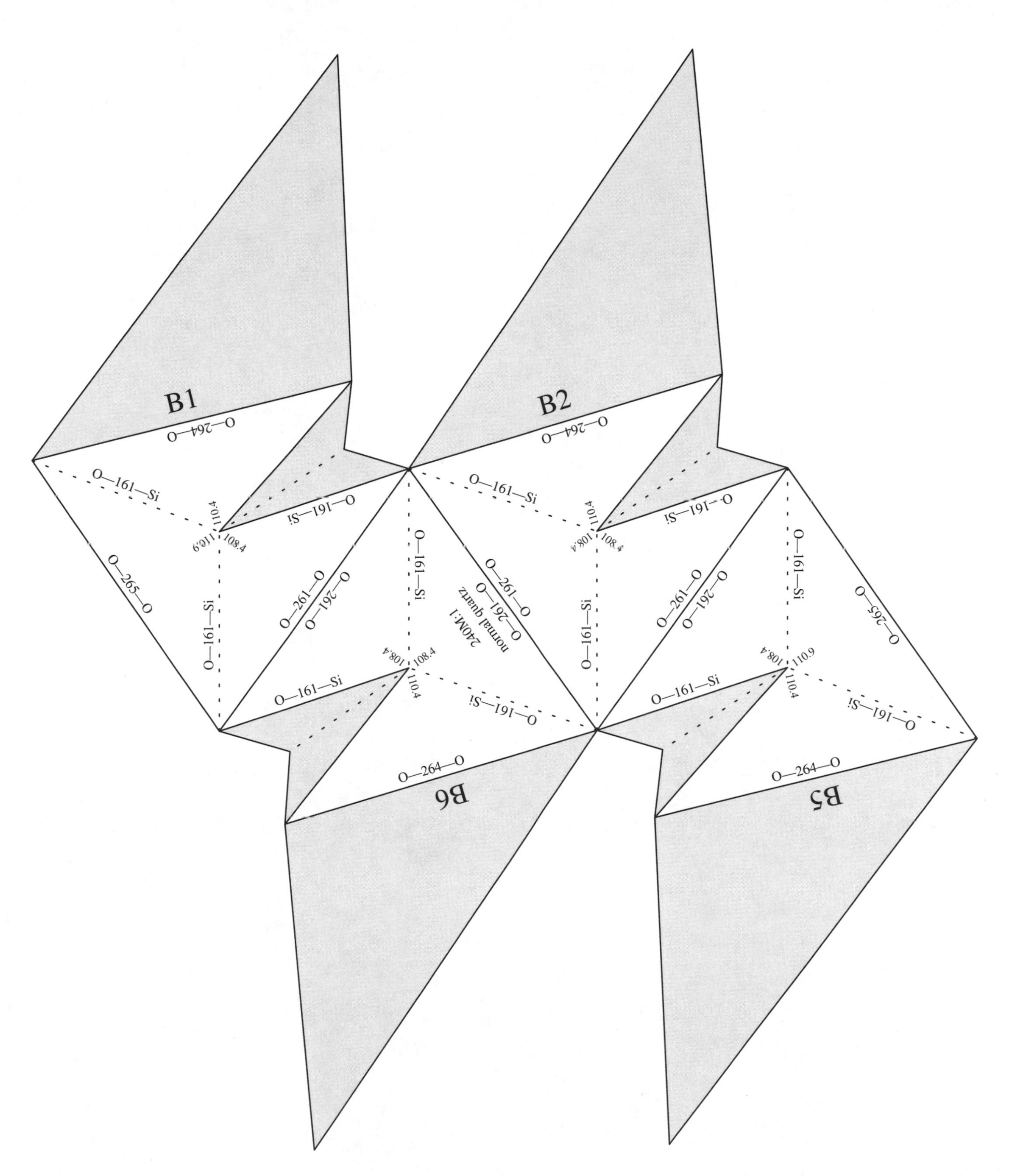

shape: corner-sharing tetrahedra units: pm scale: 240,000,000:1

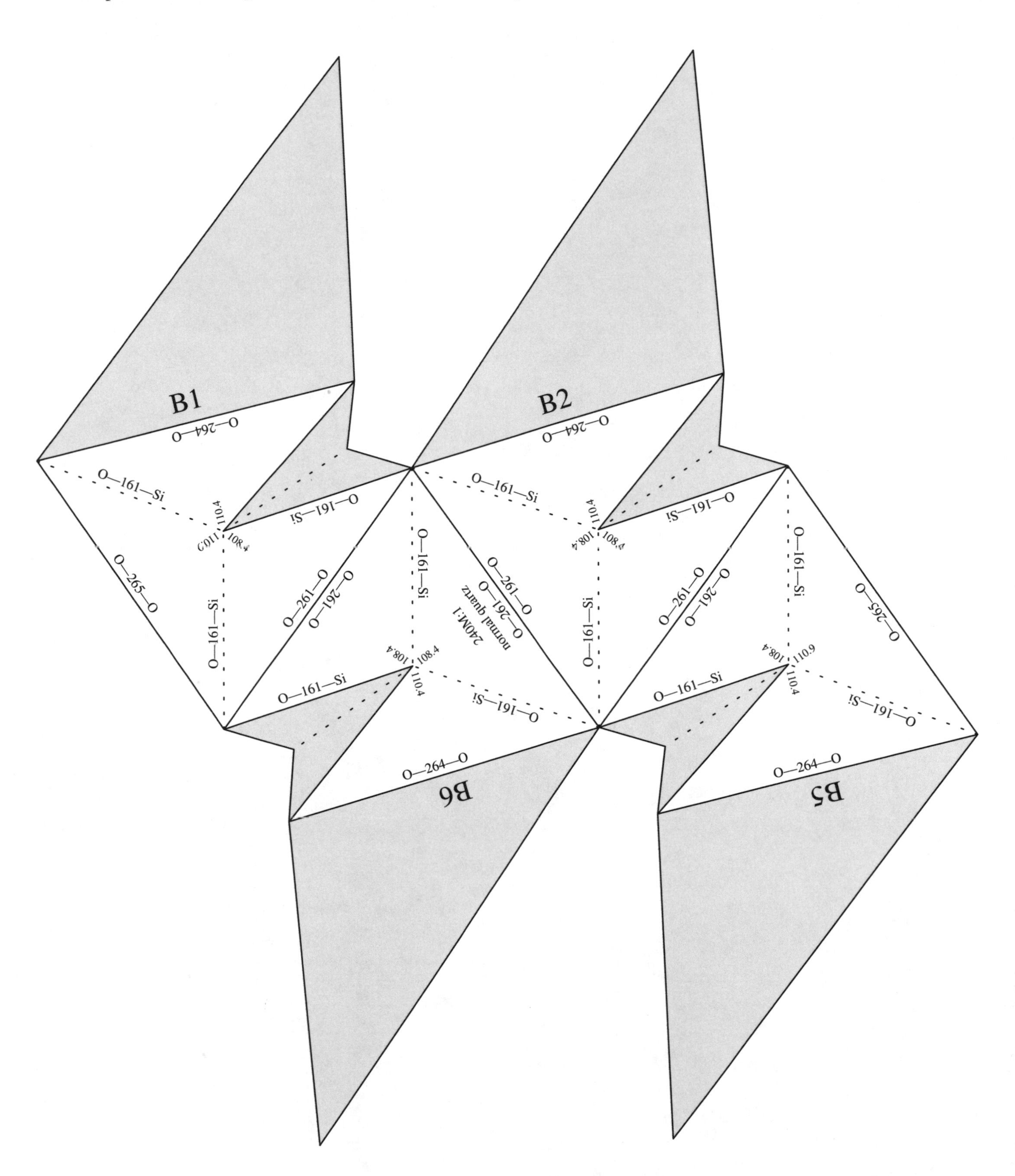

shape: corner-sharing tetrahedra units: pm scale: 240,000,000:1

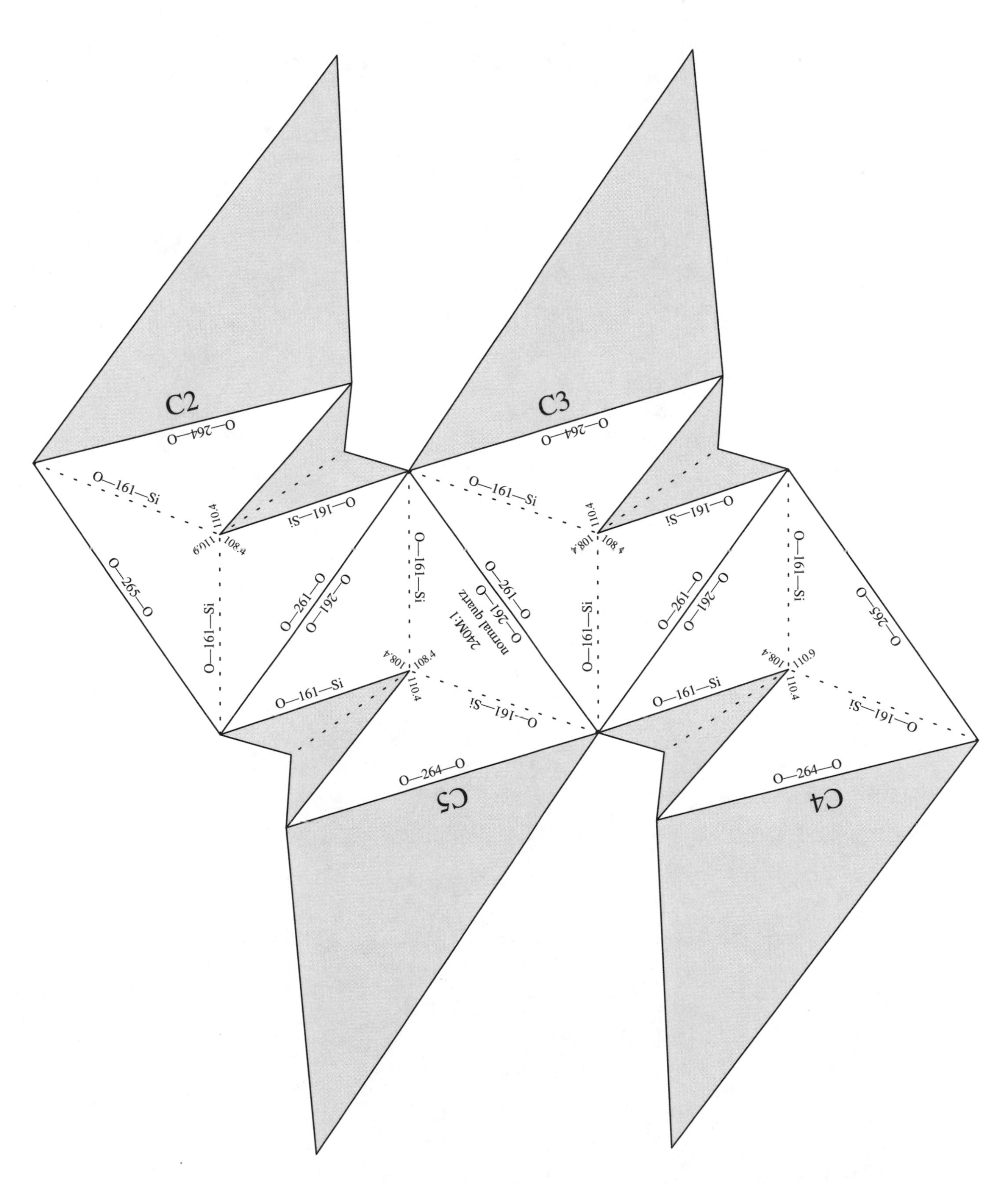

normal (α) quartz (unit C) $(SiO_2)_x$

shape: corner-sharing tetrahedra units: pm scale: 240,000,000:1

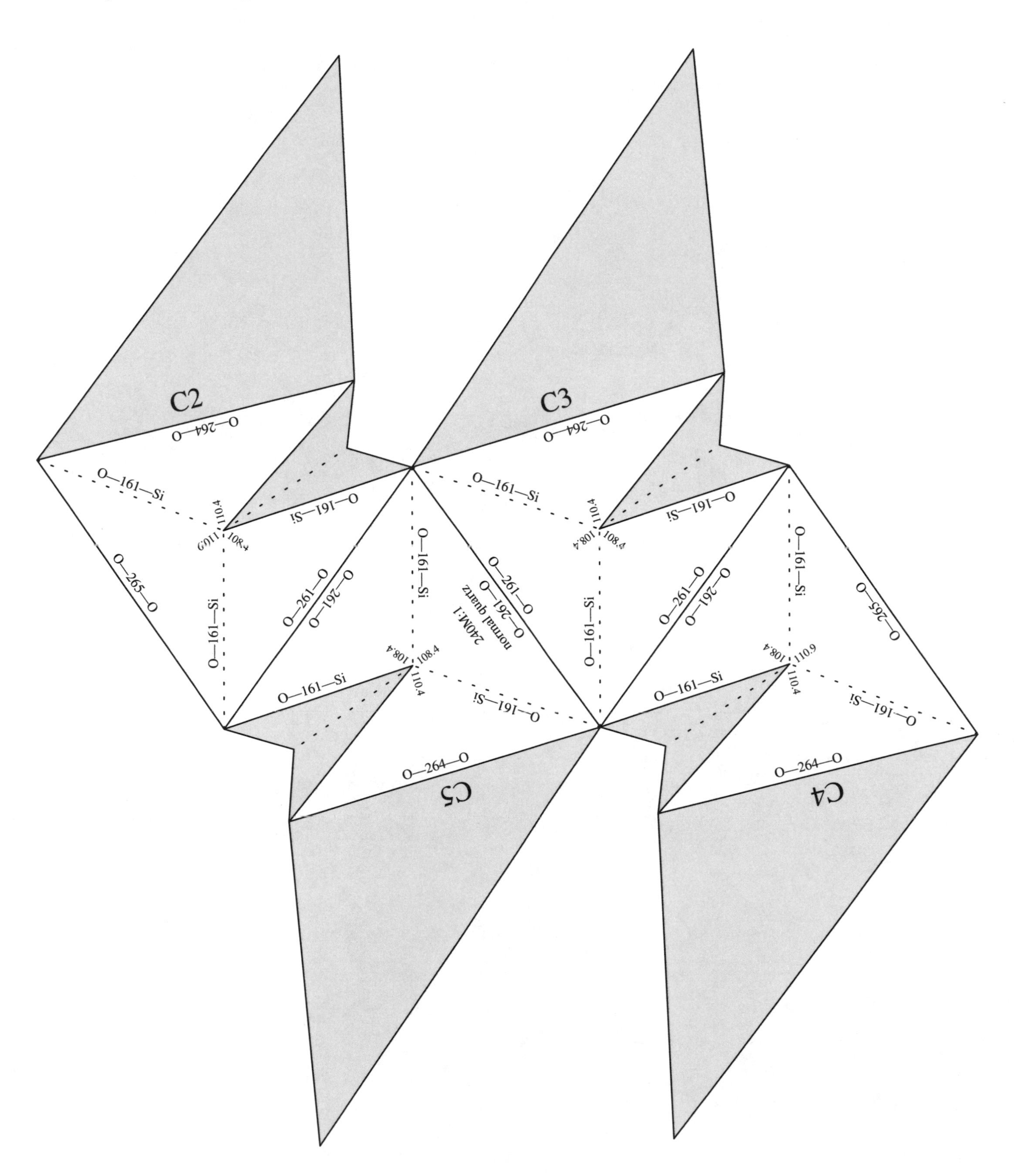

PART 6

One- and Two-Dimensional Shapes

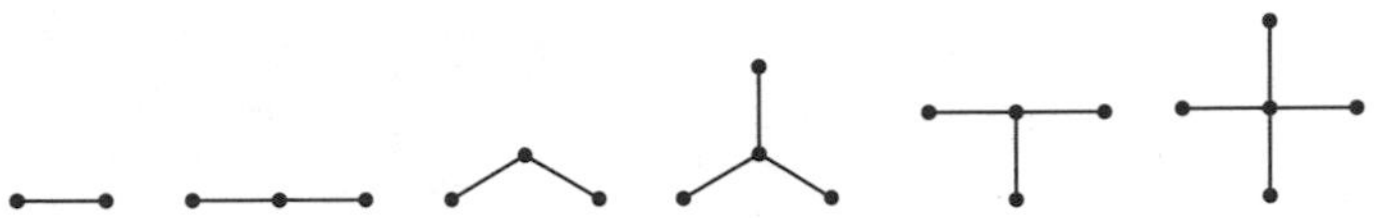

diatomic linear bent trigonal planar T-shaped square planar

In this section you will find structural data for 76 molecules and ions laid out on a scale of 300,000,000:1 (the same scale as all molecules in Part 1). In all of these cases, the arrangement of atoms *around the central atom* is either one- or two-dimensional. Not all of these molecules are flat, however. For example, the hydrogen atoms of allene are in two perpendicular planes, and the CH_3 groups of dimethyl ether are tetrahedral. This three-dimensional character has not been depicted.

The purpose of this section is simply to provide you with a reference set of data that you can use to practice predicting the shapes of molecules from formulas. It is a valuable exercise for students to write out the Lewis structures of these molecules in order to rationalize the basic shapes observed. Then, using VSEPR or molecular orbital ideas, one can address the more subtle issues of distance and angle.

For the purpose of discussing structure, the double bonds of Lewis structures are largely irrelevant. It is more useful (and easier as well) to draw "sigma" structures, having no double bonds. To make sigma structures, simply connect all atoms with single (sigma) bonds. Then count the total number of valence electrons, subtract two for every bond already drawn, and use the remaining electrons to fill out the octets of the *outer* atoms in the structure. If any electrons are left after that, give them to the central atom, but don't draw in any double bonds. The resulting electron pair distribution around the central atom determines the shape as well as the sigma "hybridization." Examples are shown below.

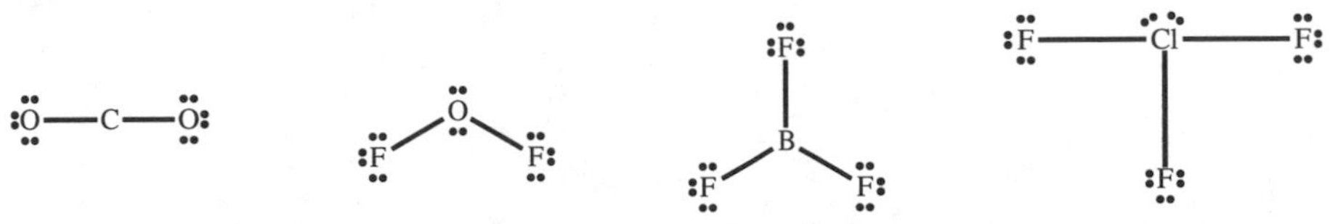

CO_2
AX_2, *sp*

OF_2
AX_2E_2, sp^3

BF_3
AX_3, sp^2

ClF_3
AX_3E_2, d^2sp^3

As a bonus, sigma structures can be used as starting points for discussions of π resonance or delocalization by keying in on the *p* orbitals left on the central atom after sigma hybridization.

units: pm scale: 300,000,000:1

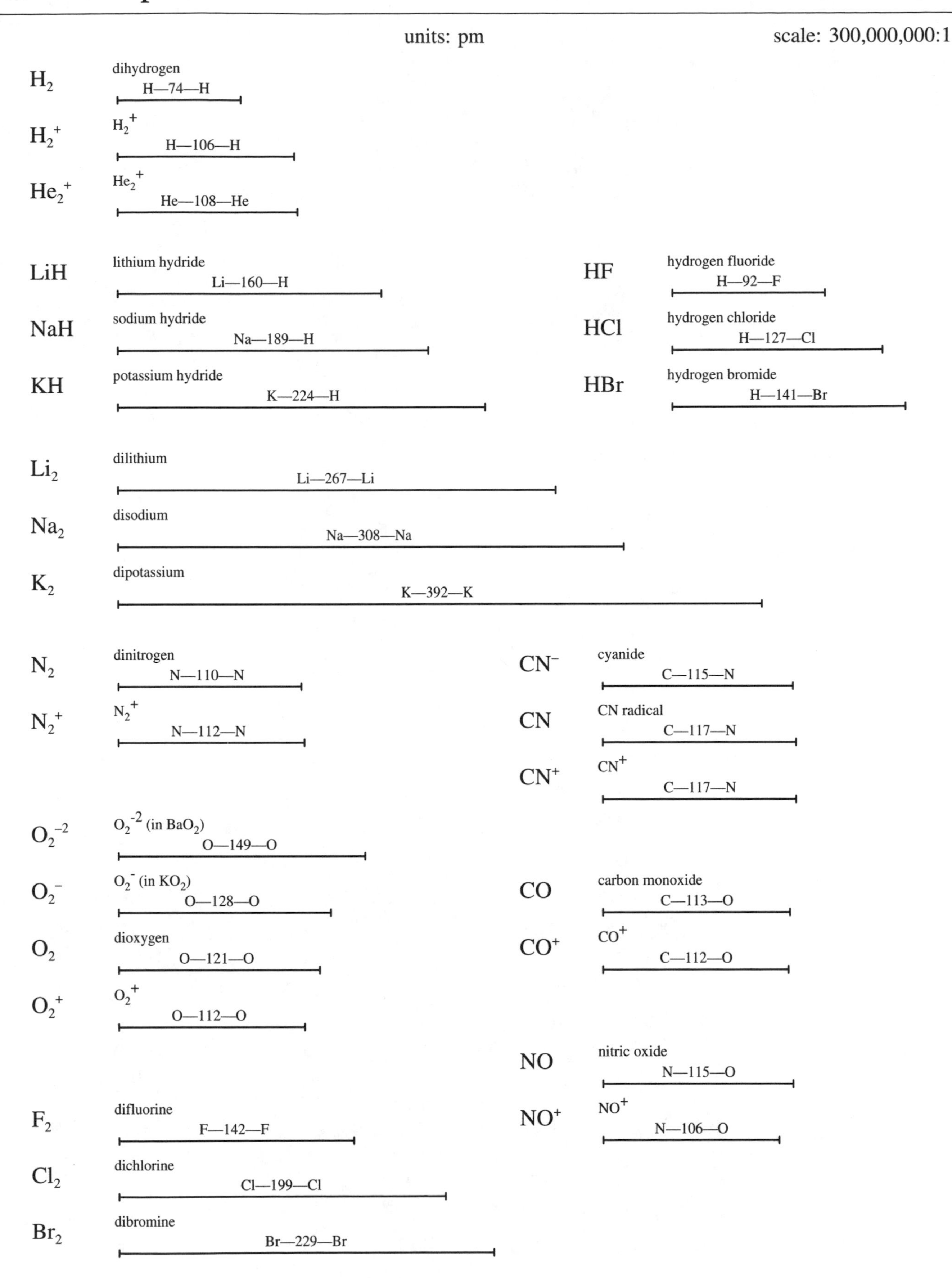

units: pm scale: 300,000,000:1

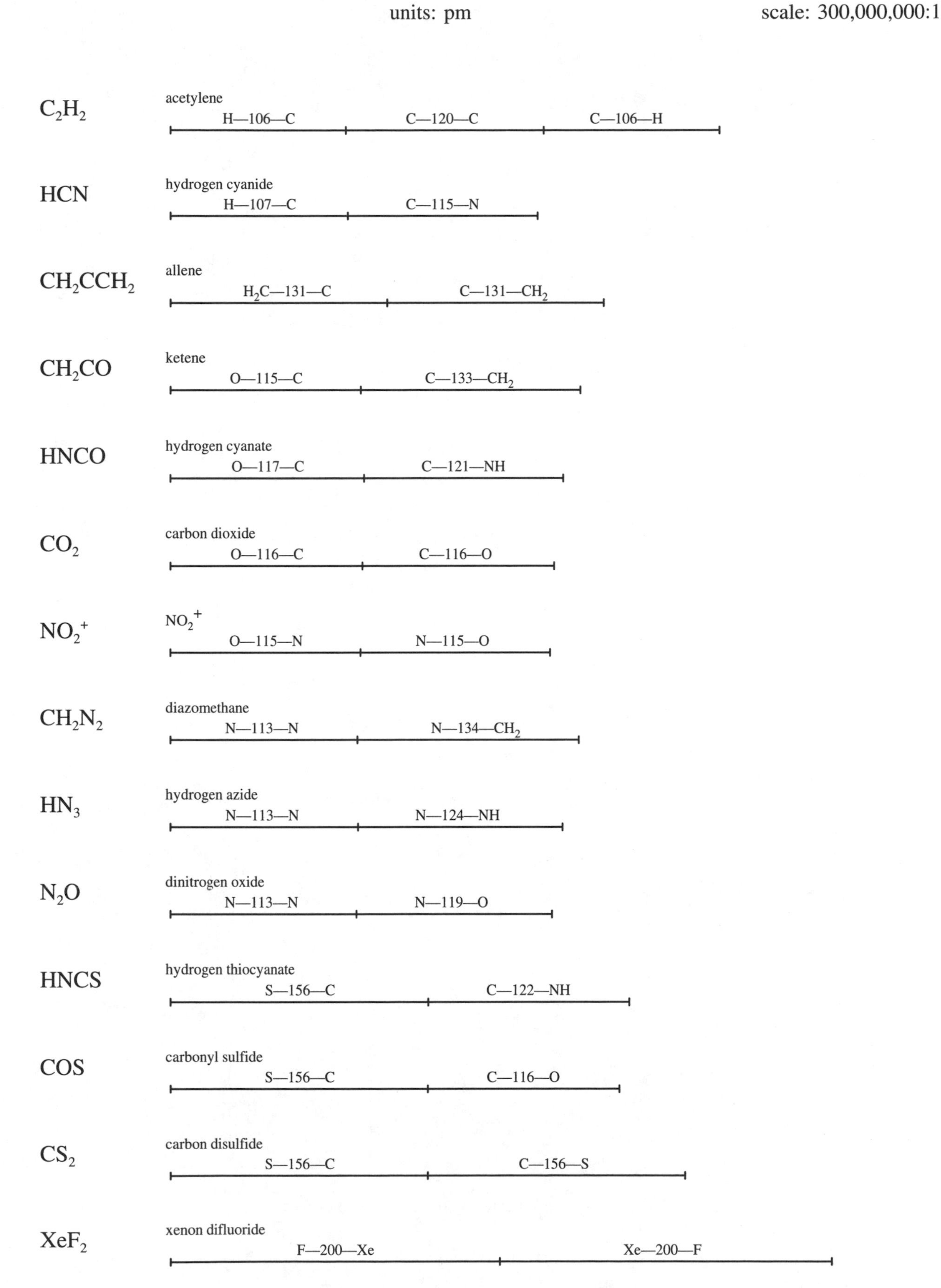

units: pm

scale: 300,000,000:1

hydrogen cyanate (N)

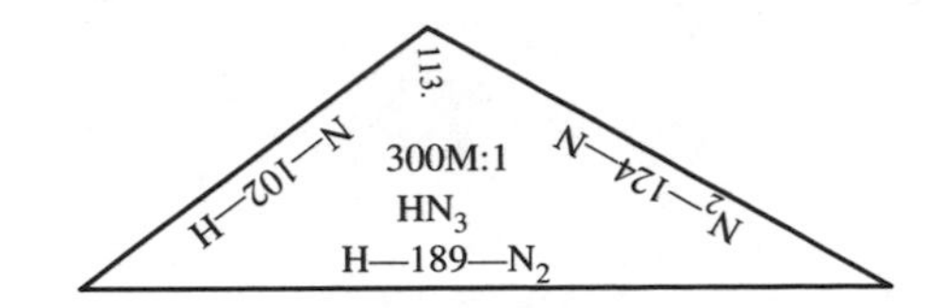

hydrogen azide (N)

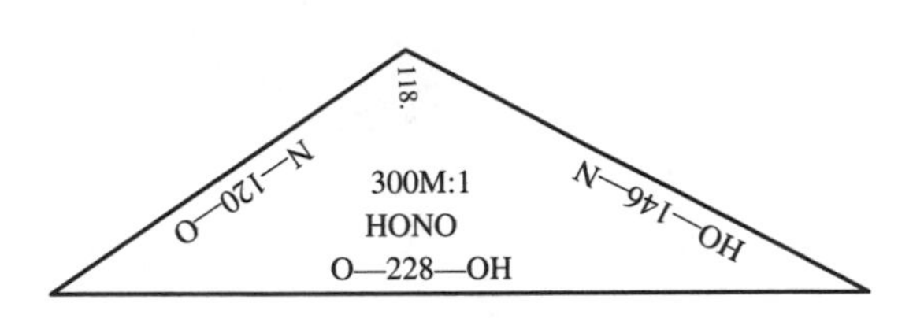

nitrous acid

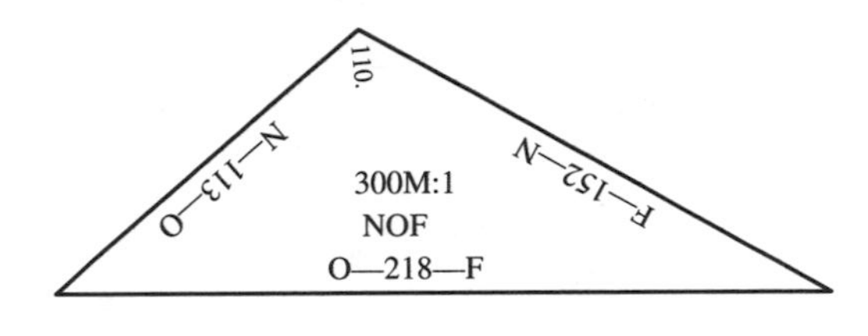

nitrosyl fluoride

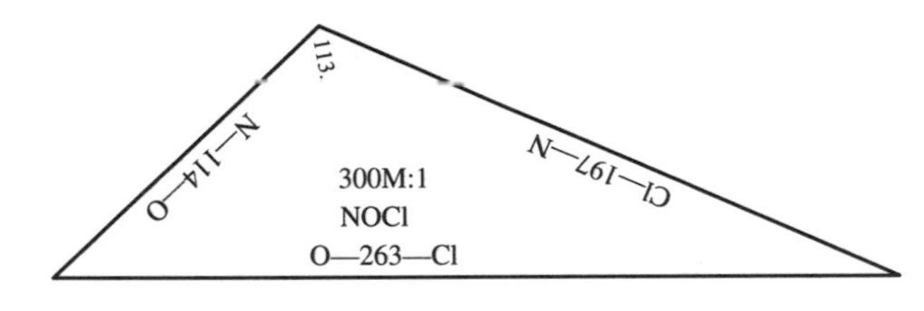

nitrosyl chloride

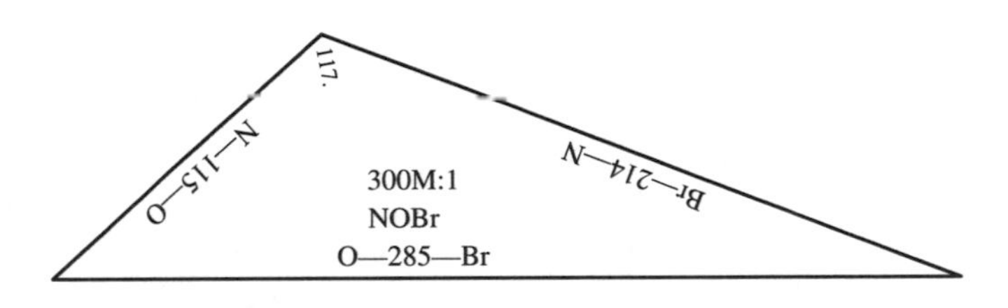

nitrosyl bromide

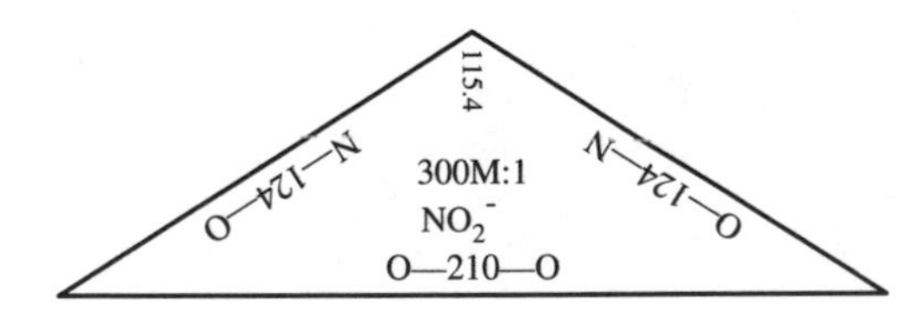

nitrite ion (in $NaNO_2$)

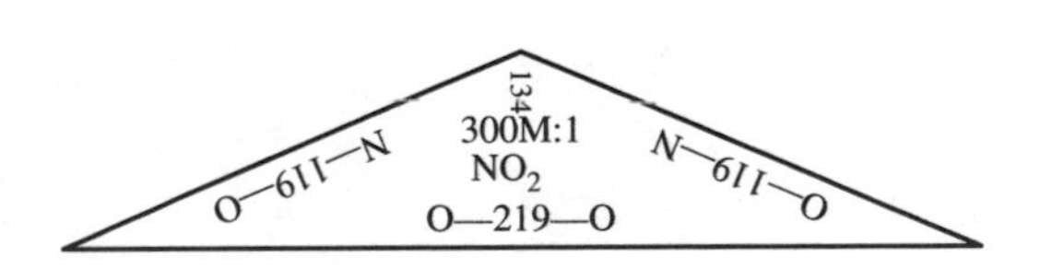

nitrogen dioxide

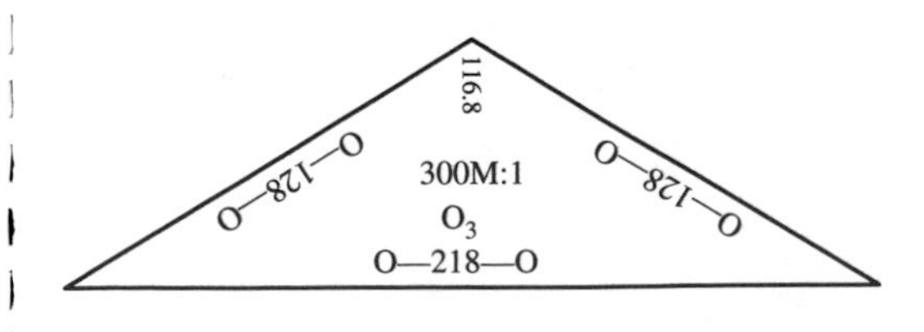

ozone

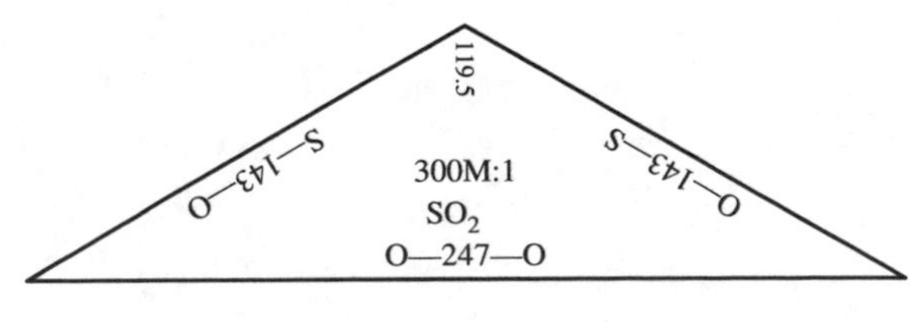

sulfur dioxide

units: pm

scale: 300,000,000:1

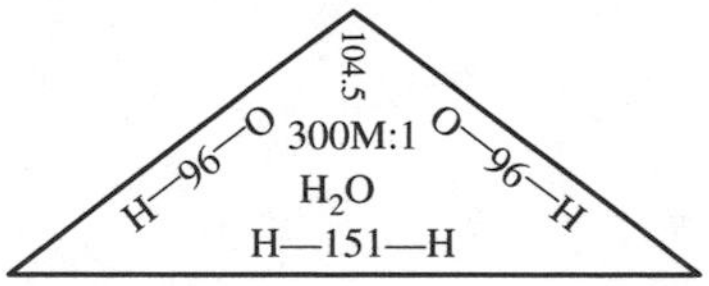

104.5

O—96—H

O—96—H

300M:1

H_2O

H—151—H

water

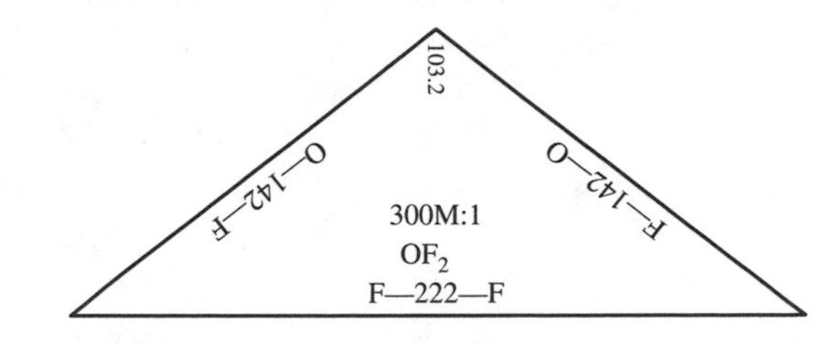

oxygen difluoride

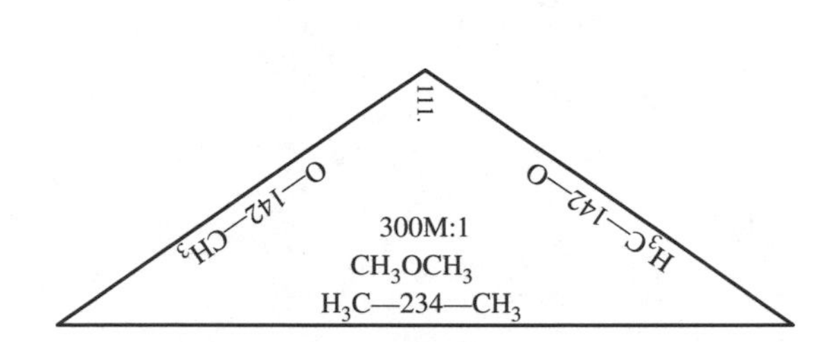

dimethyl ether

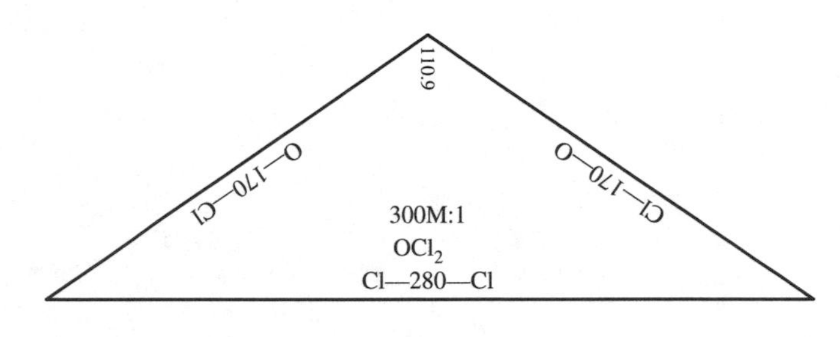

oxygen dichloride

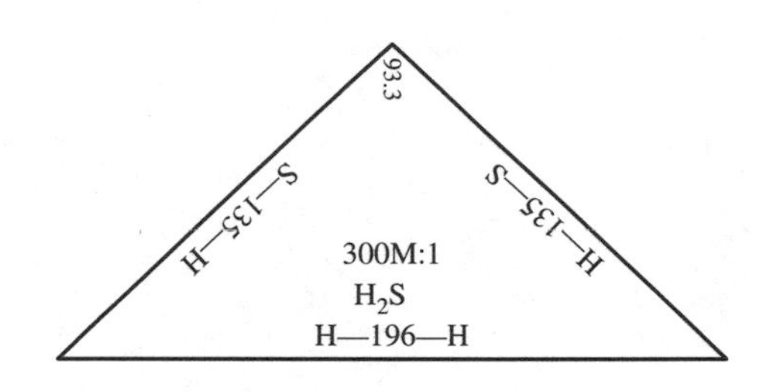

hydrogen sulfide

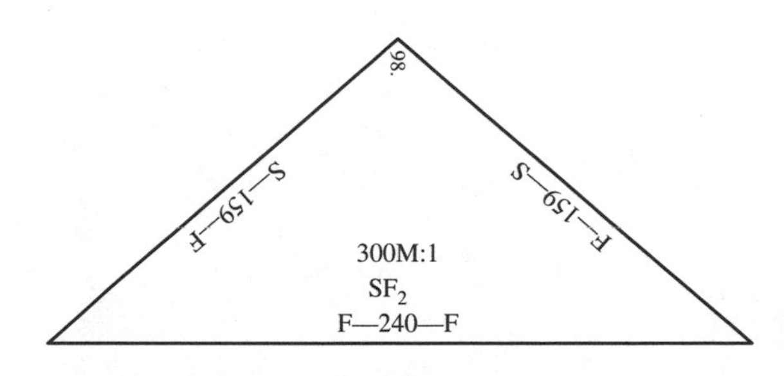

sulfur difluoride

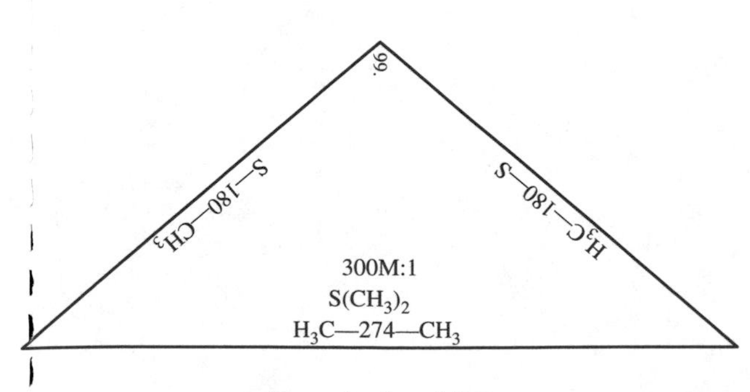

dimethyl sulfide

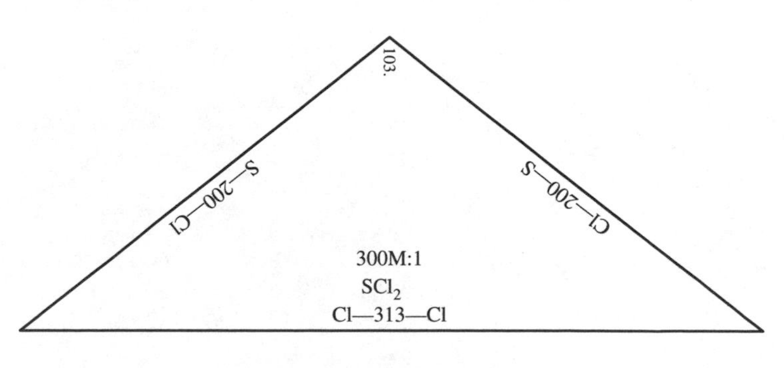

sulfur dichloride

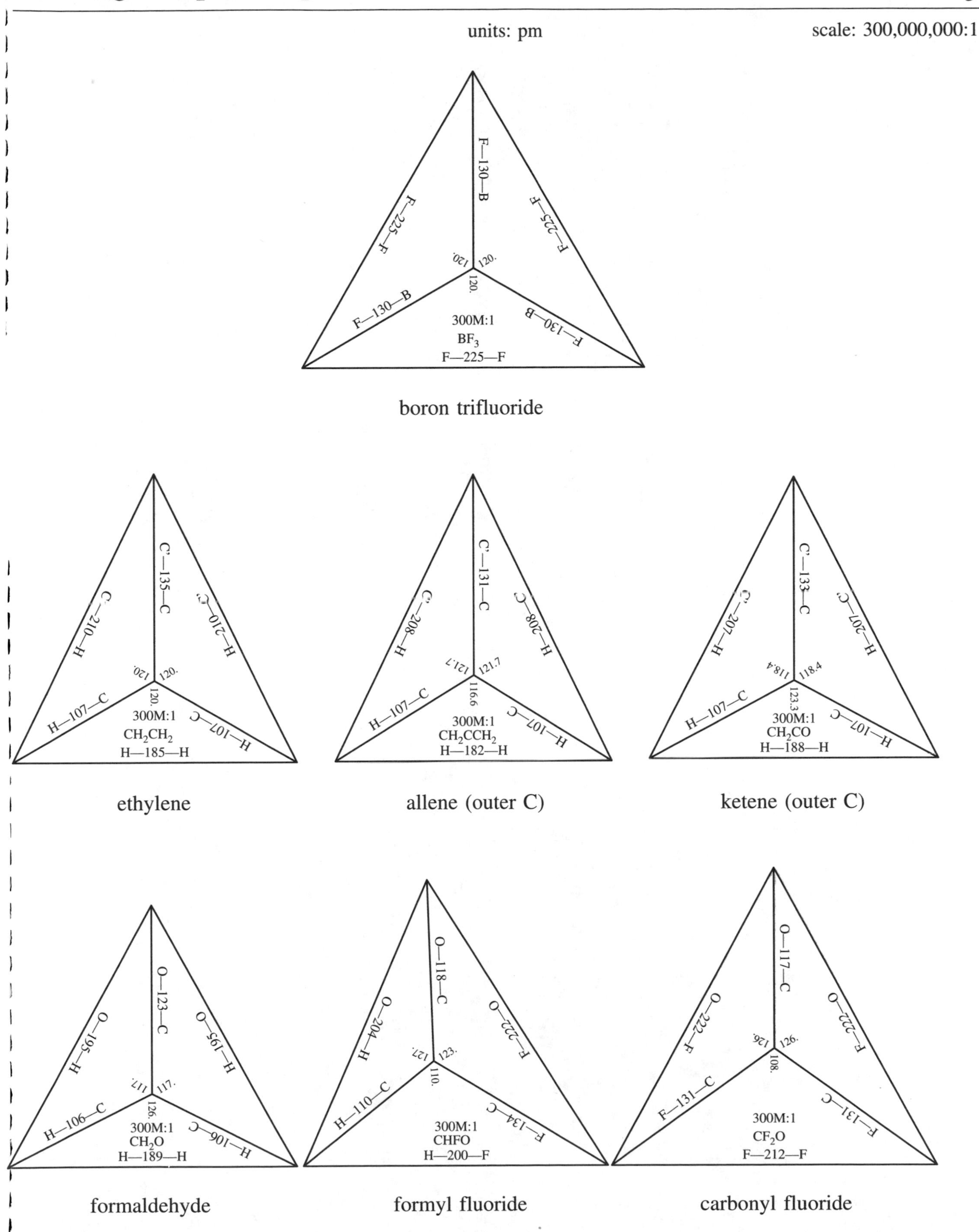
units: pm
scale: 300,000,000:1
F—130—B
F—225—F
120.
300M:1
BF3
F—225—F
boron trifluoride
C'—135—C
C'—210—H
H—107—C
120.
300M:1
CH2CH2
H—185—H
ethylene
C'—131—C
C'—208—H
H—107—C
121.7
116.6
300M:1
CH2CCH2
H—182—H
allene (outer C)
C'—133—C
C'—207—H
H—107—C
118.4
123.3
300M:1
CH2CO
H—188—H
ketene (outer C)
O—123—C
O—195—H
H—106—C
117.
126.
300M:1
CH2O
H—189—H
formaldehyde
O—118—C
O—204—H
O—222—F
H—110—C
F—134—C
127.
123.
110.
300M:1
CHFO
H—200—F
formyl fluoride
O—117—C
O—222—F
F—131—C
126.
108.
300M:1
CF2O
F—212—F
carbonyl fluoride

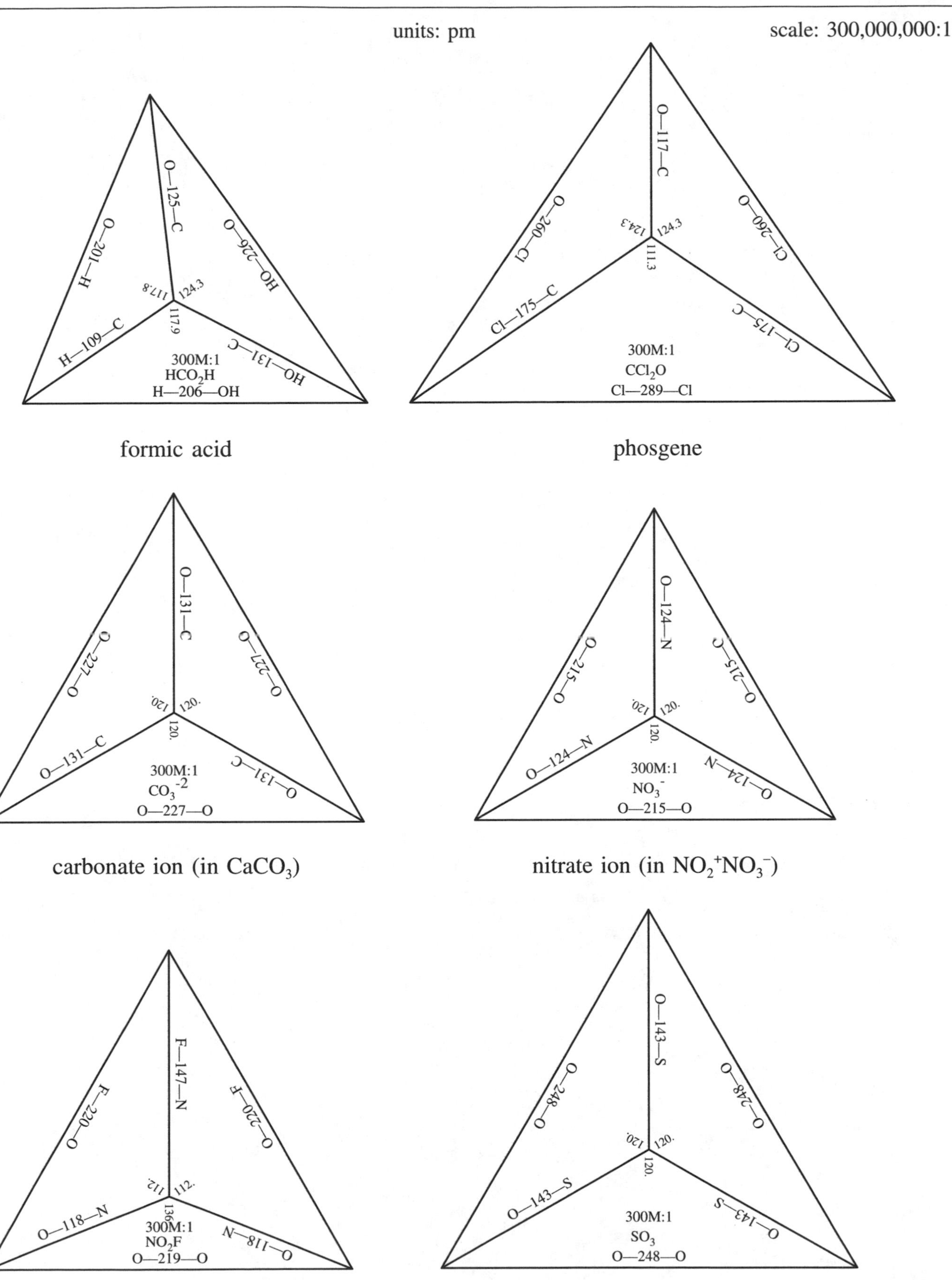

formic acid

phosgene

carbonate ion (in $CaCO_3$)

nitrate ion (in $NO_2^+NO_3^-$)

nitryl fluoride

sulfur trioxide

T-shaped planar species AX_3E_2
square planar species AX_4E_2

units: pm

scale: 300,000,000:1

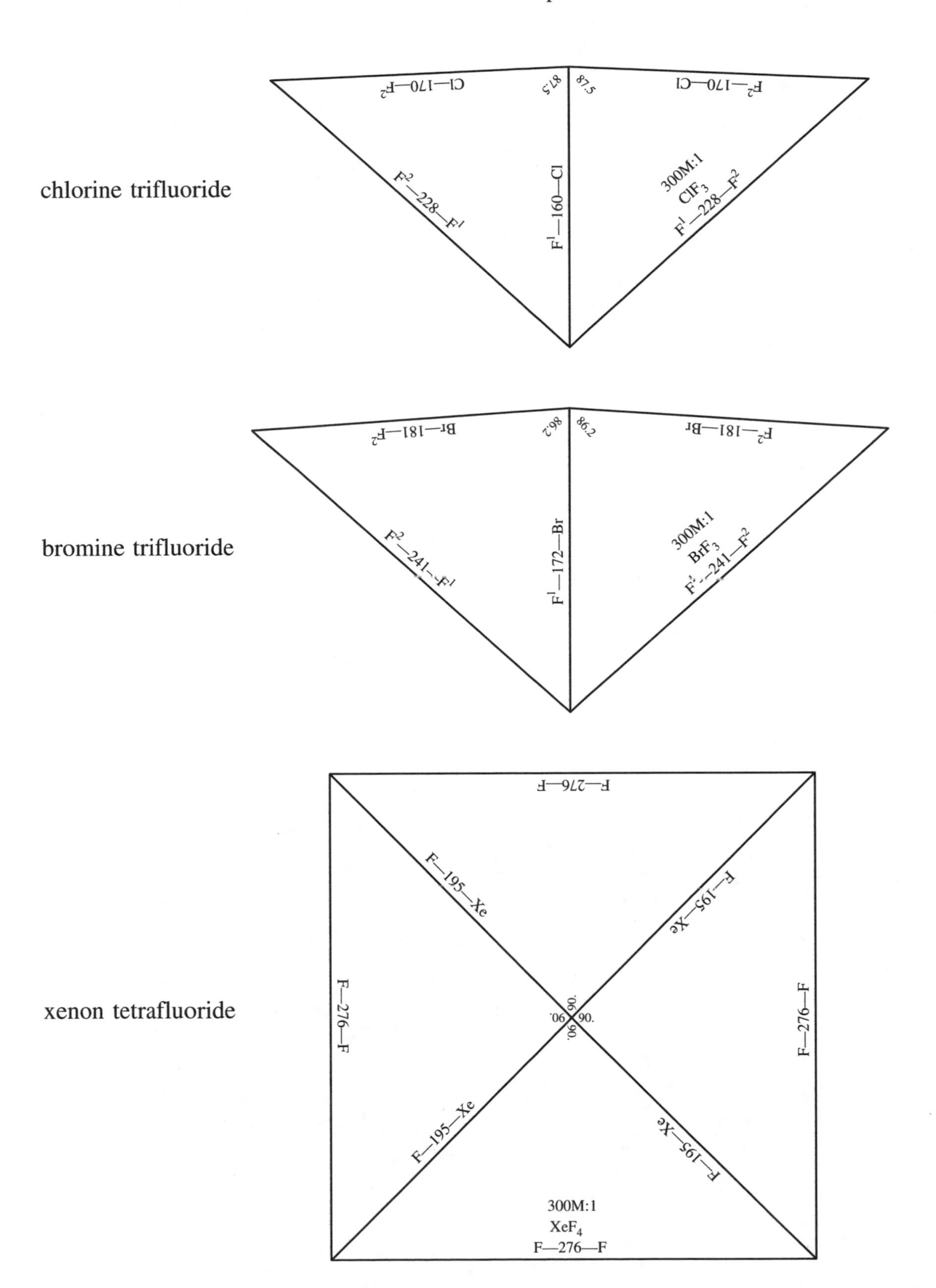

Discussion of Questions in Part 1

page 11

(a) Why isn't NH_3 flat?

In NH_3 there are, in all, eight valence electrons (five from the nitrogen and one from each hydrogen atom). Only six of these electrons are involved in bonding, the other two being a "lone pair" centered on the nitrogen. Let's focus on that lone pair and consider three possibilities for the structure of ammonia: (a) pyramidal with angles of 90°, (b) planar, with bond angles of 120°, and (c) the actual structure, with angles around 107°.

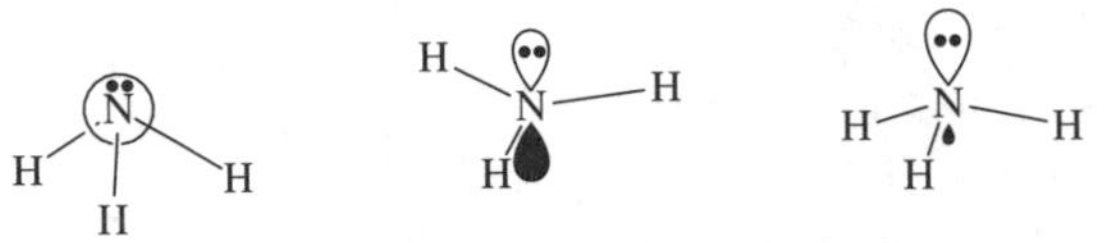

(a) pyramidal-90° (b) planar (c) pyramidal-107°

Now, if the connections were at 90° to each other, the way *p* orbitals are oriented, all of the *p* character of the central N atom would be consumed by bonding electrons, and no *p* character would be left for the lone pair. It would be pure "*s*" in character. On the other hand, if NH_3 were planar, the lone pair would have to be pure "*p*" in character, because that is the type of orbital that is left over after making three bonds in a plane. (Starting with one *s* orbital and three *p* orbitals, using three orbitals for bonding leaves one *p* orbital.) It would extend equally in both directions, above and below the plane of the atoms.

In reality, we observe something in between these two extremes, and we say that the lone-pair electrons (as well as the bonding electrons) are in "hybrid" orbitals that have both *s* and *p* character. Essentially, the angle of 107° is evidence that the lone pair has some *p* character to it, giving it *direction*. What we are seeing is the effect of a molecular energy balancing act. Atomic *s* orbitals are lower in energy than *p* orbitals, so a lone pair with more *s* character is a lower-energy lone pair. But if the lone pair has *all* of the *s* character—as in (a)—then the electrons in the *bonds* are destabilized, because they have none. In contrast, giving all of the *s* character to the bonds—as in (b)—is even worse, because then the lone pair is left highly destabilized. What we see is intermediate between these two extremes.

Overall, situation (c) is the best of all worlds, with both the lone pair and the bonding electrons in "hybrid" orbitals. All four atomic orbitals on nitrogen are used for holding electrons in bonds or lone pairs. We say that NH_3 is "sp^3" hybridized.

> (b) Both CH_4 and NH_3 consist of 10 protons, 10 electrons, and several neutrons. Compare the structure of NH_3 with that of CH_4 (page 31). How are they similar? How are they different?

It is interesting to compare the structures of CH_4 and NH_3 because the two have exactly the same number of protons and electrons. CH_4 is *larger* than NH_3, and the H–C–H angles in CH_4 are larger than the H–N–H angles in NH_3. That is, the NH_3 molecule is a smaller, steeper pyramid than the CH_3 part of CH_4. (Make both models and see if you can't hide the NH_3 model under the CH_4 model with room to spare.) What might be the cause of this?

> (c) Imagine "magically" turning CH_4 into NH_3 by moving one of the protons we call "H" into the carbon nucleus to form "N." How are the structural differences between CH_4 and NH_3 consistent with this transformation, given that protons attract electrons?

The idea of moving a proton from one nucleus to another, in effect imagining "nuclear" chemistry, is what I call "proton relocation." Nobody can really do this, but it's fun to think about trying. In CH_4 there are no central atom lone pairs, and all eight valence electrons are involved in bonding—in tying together the central atom with the four protons. (Remember, in molecules, "H" is just a proton.) The proton relocation going from CH_4 to NH_3 has two major effects, as illustrated below.

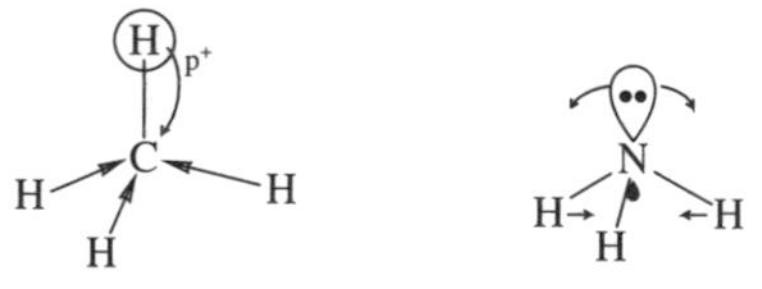

methane   ammonia

- The relocated proton is now more central, more able to attract *all* of the electrons in the molecule. Relocating a proton into a central position is likely to make a molecule smaller.
- The two electrons now left out in the cold (more like a game of musical chairs, really, where one of the chairs, an orbital, has suddenly disappeared) are especially affected. Those electrons used to be stabilized by that proton we called an H atom, and they still need stabilization. They get it by taking on more low-energy central atom *s* character *at the expense of the other electrons.*

The first of these effects explains the smaller size of NH_3 relative to CH_4. The second effect explains the difference in angles. Remember, this is a "zero-sum" game in terms of divvying up the *s* and *p* character of the molecular orbitals. In going from CH_4 to NH_3, the lone pair gains *s* character in order to "follow" the proton toward the central nucleus. That *s* character must come from the six electrons not involved in the lone pair (the remaining six *bonding* electrons). They have *lost s* character. The bonding electrons are still "sp^3," but now they have a little less "*s*" and a little more "*p*" than they had before the relocation.

(d) Imagine magically turning NH_3 into H_2O the same way. What do you predict for the structure of the water molecule?

The pattern for H_2O, shown also at a scale of 300,000,000:1, is shown on the right. The X–H distance in H_2O is smaller than in NH_3. The H–X–H angle in H_2O is smaller as well. The distance change is easily rationalized by proton relocation. But what about the angle change? Is that predictable? I think not.

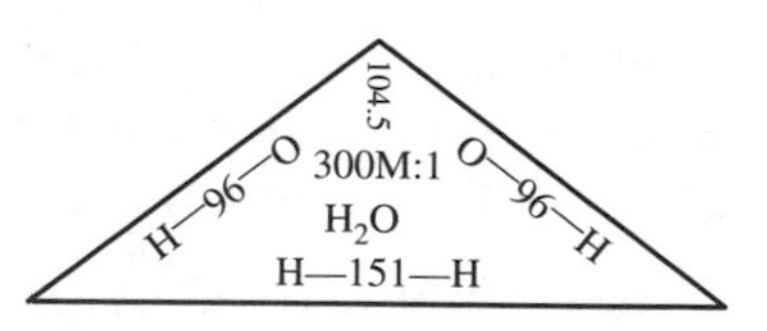

In H_2O there are two lone pairs, and nothing in the world says they have to be identical. Calculations and experiment both indicate that one of the lone pairs in H_2O is high in energy and pure *p* in character. That leaves the other electrons in sp^2 orbitals (leading, ideally, to bond angles of 120º). Thus, based on this information, we might expect the H–O–H angle in H_2O to be *larger* than the H–N–H angle in NH_3 (106.6º). But there is that lone pair in the plane of the hydrogen atoms. It should take on extra *s* character and drive the H–X–H angle smaller. Should it end up smaller than 106.6º? Who can say? As it turns out, the angle in water (104.45º) is just a bit smaller than the angle in NH_3. We probably shouldn't make too much of only a 2º difference.

The trend of decreasing angle as one goes from CH_4 to NH_3 to H_2O is often attributed to the angular "size" of lone pairs being larger in some way than the angular "size" of an O–H bond. But that is debatable. In our terms, this larger angular size simply translates to having more *s* character. But once there are two lone pairs in a molecule, it is unclear to what extent one will take on *s* character at the expense of the other electrons in the system. It is quite possible that this "trend" is due to much more subtle electronic factors that should not be oversimplified except as a mnemonic tool for remembering some of the facts of nature. We'll see in the next examples that all is not so simple.

(e) Ammonia and water react as follows to form the ammonium ion (NH_4^+) and hydroxide (OH^-):

$$NH_3 + H_2O \longrightarrow NH_4^+ + OH^-$$

What do you predict (shape, angles, and distances) for the structure of NH_4^+ (page 35)?

Adding an outer proton to NH_3 gives ammonium, NH_4^+. It should be the same basic shape as CH_4, tetrahedral. The angles in NH_4^+ should also be the same as those in CH_4. In terms of size, NH_4^+ should be larger than NH_3 (but still smaller than CH_4). That's because the extra proton not only interacts with the lone pair, but also with *all* of the electrons in the molecule, effectively making each N–H connection a little looser than the ones in NH_3. But CH_4 has one less proton in its center than does NH_4^+, so CH_4 should be larger still.

(f) Ammonia and boron trifluoride (BF_3, which is flat) react to form $BF_3{\cdot}NH_3$. What do you predict for the structure of $BF_3{\cdot}NH_3$ (page 41)?

Reaction of NH_3 with BF_3 leads to a complex preserving the pyramidal nature of the NH_3 unit (now tetrahedral). The exact H–N–H angle in $NH_3{\cdot}BF_3$ is not known, but calculations indicate that it should be around 107º (Hehre, p. 216). It is also interesting to compare the B–F distance in $BF_3{\cdot}NH_3$ to the B–F distance in BF_3.

page 13

(a) How does the structure of NF_3 compare to that of NH_3 (page 11)?

A comparison of the structures of NH_3 and NF_3 teaches us the effect of orbital size and atomic *electronegativity*, the tendency of an atom to hold onto shared electrons. NF_3 is larger, with tighter angles than NH_3. The larger size has to do with two factors:

- The 2*p* orbitals used by fluorine are intrinsically larger than the 1*s* orbitals used by hydrogen.
- The six lone pairs on each fluorine cause problems, in that they strongly repel the lone pair on nitrogen. Thus, electron-electron repulsion tends to keep the N and F more at a distance.

The tighter angle in NF_3 indicates extra *p* character for the bonding electrons in that molecule. Once again, for an explanation we turn to the lone pair. How does the presence of fluorine atoms affect the lone pair? Fluorine is more *electronegative* than hydrogen, meaning that when sharing electrons (as in NF_3 and NH_3), F holds onto them more tightly than does H. The effect of that increased electronegativity is for all of the bonding electrons in the entire molecule to spend more time closer to F and farther from N. As a result, the nitrogen atom has a higher *effective nuclear charge* in NF_3 than it has in NH_3.

The lone pair on nitrogen is most attracted to this charge and takes advantage of it by gaining *s* character. That leaves *less s* character for the bonding electrons in NF_3. They are left with proportionately more nitrogen orbital *p* character, making the X–N–X angle closer to 90°.

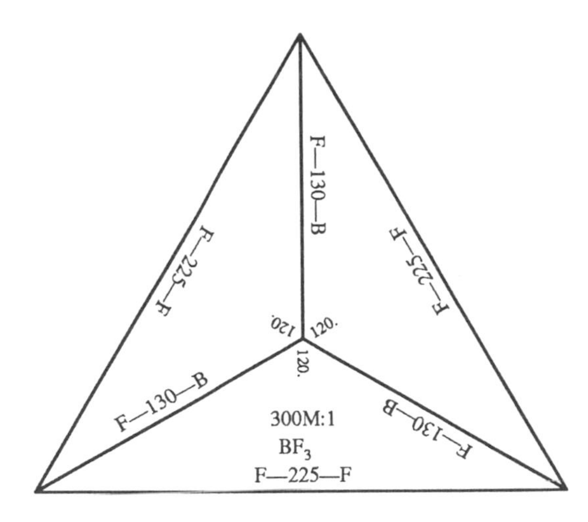

(b) Compare the structure of NF_3 to that of BF_3, shown to the same scale on the left. What are the differences? Explain them.

The crucial difference between NF_3 and BF_3 is the presence or absence of a central atom lone pair. BF_3 is smaller than NF_3 and is flat instead of pyramidal. Both differences are due to the fact that there are only six valence electrons in BF_3. All six are involved in bonding, and none are left in the form of a lone pair. Consider again the same three possibilities:

(a) planar (b) pyramidal-90° (c) pyramidal >90°

With no central lone pair, there is nothing to stabilize except the bonding electrons. They take all of the *s* character, and situation (a) is observed. In situations (b) and (c), *s* character is being "wasted" on empty orbitals. An extra bonus for BF_3 is that the empty *p* orbital can actually still participate in electron stabilization. Some of the orbitals on fluorine can combine with that space to form a *delocalized pi bond*:

Though weak, this pi interaction probably shortens the B–F distance even further. Notice, too, that with no central atom lone pair in BF_3 there is not the lone pair e^--e^- repulsion present in NF_3.

(c) What would you predict for the distances and angles in CHF_3 (page 57)?

In CHF_3 we have relocated a proton out of the central nucleus of NF_3, tying up the lone pair. The effect of this is threefold:

- Pulling a proton out of the central nucleus diminishes its pull on all of the electrons in the molecule, tending to make the distances *larger*.
- On the other hand, tying up the lone pair in a bond removes the problem of lone pair e^--e^- repulsion and might result in distances being *smaller*.
- Finally, tying up the lone pair gives lone-pair *s* character back to the other bonding electrons, tending to make the angles less "*p*-like"—less perpendicular—that is, *larger*.

Now, the first two of these effects counteract one another. However, it turns out that the second, electron-electron repulsion, is very important with fluorine *specifically* because fluorine atoms are so small and so packed with electrons. The result is that CHF_3 has smaller X–F distances and larger F–X–F angles than NF_3. The differences are depicted below:

(d) In CF_4 (page 33), the C–F distances are all 132 pm and the angles are all 109.5°. The structure of OF_2 is shown on the right. Is there a trend here as there is for the series CH_4, NH_3, H_2O?

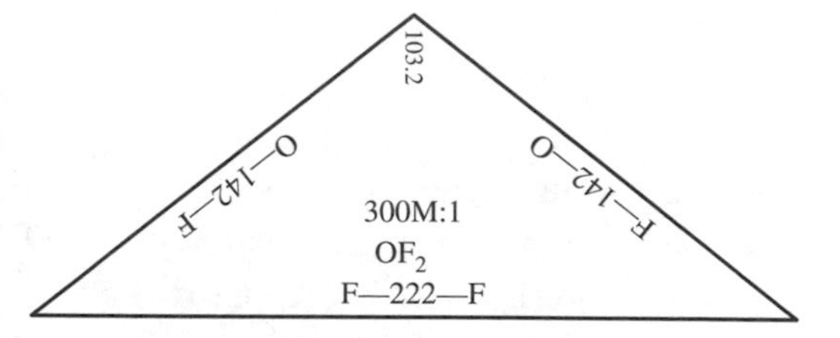

If the series CF_4 to NF_3 to OF_2 represented a meaningful trend, one might expect the O–F distance in OF_2 to be even larger than the N–F distance of 137 pm in NF_3 and the F–O–F angle to be even smaller than the F–N–F angle of 102.5°. The distance increase is borne out; the angle decrease is not. But, then, is this a meaningful comparison? This series is not just like the series CH_4 to NH_3 to H_2O. Those molecules all have the same number of protons and electrons. In going from CF_4 to NF_3 to OF_2, all bets are off. The differences between a fluorine atom and a hydrogen atom are like night and day.

page 15

(a) What are the main differences between the structure of NCl_3 and those of NH_3 (page 11) and NF_3 (page 13)?

NCl_3 is much larger than either NH_3 or NF_3. Its angles are about the same as in NH_3 and considerably larger than those of NF_3.

(b) Why are the distances in NCl_3 so much larger than in NF_3?

The larger size of NCl_3 relative to NF_3 is attributable to chlorine being in the second row of the periodic table while fluorine is in the first. The valence electrons of chlorine (3*s* and 3*p*) are much farther from the nucleus than those of fluorine (2*s* and 2*p*). This *always* translates into larger interatomic distances for chlorine than for fluorine.

(c) Why do you think the angles in NCl_3 are almost the same as in NH_3?

The angle similarity between NCl_3 and NH_3 is probably completely fortuitous, an effect of many factors. More than anything, there just isn't as much of a problem of lone pair e^--e^- repulsion when one of the lone pairs is on a second-row atom such as chlorine.

(d) What do you predict for the structure of $CHCl_3$ (page 59)?

In $CHCl_3$ we have relocated a proton out of the central nucleus of NCl_3, tying up the lone pair. This situation is analogous to the comparison between CH_4 and NH_3. The distances and angles in $CHCl_3$ are larger than they are in NCl_3, just as the distances and angles in CH_4 are larger than they are in NH_3. Note that chlorine does not seem to show the same problem of lone pair e^--e^- repulsion that is seen for fluorine. Fluorine is truly a unique element. A tremendous amount of its chemistry seems to derive from its being so small and so electron-rich. Chlorine does not exhibit the same chemistry as fluorine because the 3*p* orbitals in chlorine are much larger than the 2*p* orbitals in fluorine.

(e) Shown to the right is the structure of OCl_2. Compare this structure to that of NCl_3 and predict the distances in CCl_4.

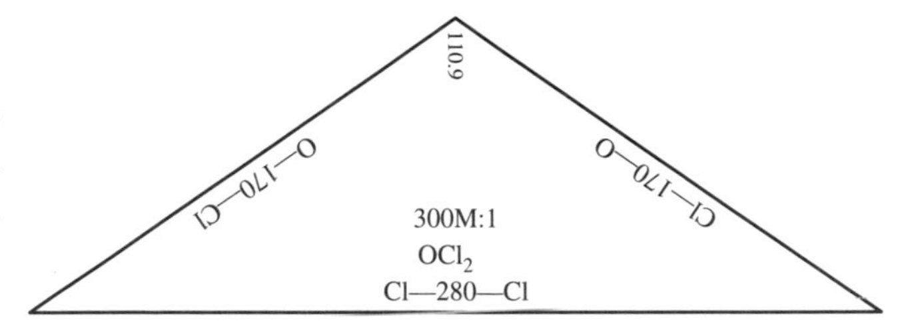

The series OCl_2, NCl_3, CCl_4 is meaningful only in relation to interatomic distance (since the angle in CCl_4 is fixed at 109.47° by symmetry). Based on this information, one would predict that the C–Cl distance in CCl_4 would be larger than 175 pm. (It is, in fact, 177 pm.) But are these really meaningful comparisons, in light of the effects of lone pairs, or is this just a fortuitous trend?

page 17

(a) How does the structure of NH_2CH_3 compare to that of NH_3 (page 11)? How do you explain the differences?

The H–N–H angle in NH_2CH_3 is a little smaller than the corresponding angle in NH_3 (105.8° vs. 106.6°). The CH_3–N–H angle is much larger (112.2°). More than anything, this is probably due to the fact that CH_3 is much larger than H.

(b) Compare the structure of NHF_2 to that of OF_2 (see page 13). How do you explain the differences in X–F distance?

Relocating a proton in NHF_2 into the central atom to give OF_2 produces a new lone pair. Based on our observations with NF_3, this might cause problems and lead to a *larger* X–F distance in OF_2 relative to NHF_2. This is, in fact, the case.

(c) Imagine pulling one of the protons in the N out of each of these molecules to form CH_3CH_3 (page 49) and CH_2F_2. What do you predict for the structures of CH_3CH_3 and CH_2F_2?

Relocating a proton of NH_2CH_3 out of a central nucleus to give CH_3CH_3 should show the same effect as going from NH_3 to CH_4: larger H–X distances (no lone-pair–lone-pair effect here) and a more even distribution of *s* character (more tetrahedral angles).

As for NHF_2 and CH_2F_2, here is our lone-pair–lone-pair effect again. The effect should be the same as that observed for NF_3 and CHF_3. That is, the C–F distance in CH_2F_2 is likely to be *smaller* than the N–F distance in NHF_2 (140 pm) because of the absence of a central lone pair in CH_2F_2. Sure enough, it's 136 pm.

page 19

(a) Phosphorus is just below nitrogen on the periodic table. How does the structure of PH_3 compare in shape, distance, and angle to that of NH_3?

PH_3 has the same basic pyramidal shape as NH_3; however, the distances in PH_3 are larger and the angles are smaller.

(b) What do angles near 90° mean in molecular structure?

As mentioned in the discussion of question (a) on page 11 (see page 191), if three atoms are attached to a central atom using only *p* orbitals, the three central connections would be perpendicular to each other, just the way *p* oribitals are perpendicular. Thus, angles near 90° in PH_3 mean lots of *p* character for the bonding electrons in that molecule.

(c) Shown to the right is the structure of H_2S shown to the same scale. How do you explain the differences in structure between H_2S and PH_3?

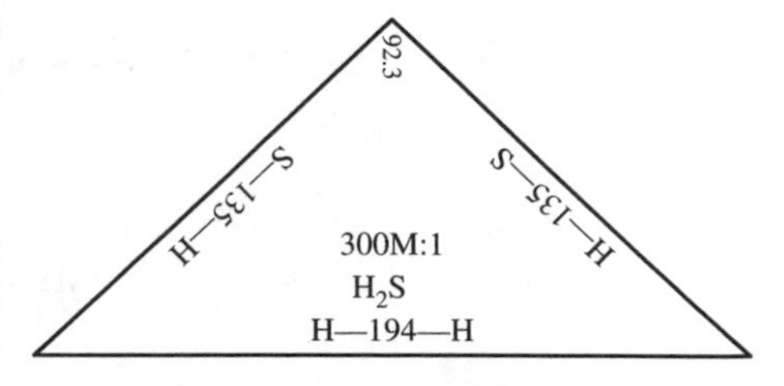

Like the differences between H_2O and NH_3, those between H_2S and PH_3 can be accounted for based on proton relocation. The smaller size of H_2S can be attributed to the stronger central nuclear attraction for electrons. As for angle, the difference is probably so slight as to be meaningless. Apparently, little rehybridization is required. (Remember, the lone pair in PH_3 already has virtually all of the *s* character available in the system.)

(d) What would you predict for the shape, distances, and angles of SiH_4 (page 61)?

In contrast to the significant differences in shape between PH_3 and NH_3, SiH_4 is an ideal tetrahedron, a larger version of CH_4, with angles of 109.47°. SiH_4 should also be larger than PH_3 based on proton relocation. Indeed, the Si–H distance in SiH_4 is 154 pm, compared to a P–H distance of 142 pm in PH_3.

(e) Why might PH_3 be less basic than NH_3?

PH_3 is almost completely nonbasic ($pK_b \approx 26$). The main reason has to do with *rehybridization*. The 3*s* orbital on phosphorus used by the lone pair in PH_3 is quite low in energy. In order for that lone pair to react with an acid, the electrons must be dragged out of their *s*-character depths. It just doesn't happen. In contrast, rehybridization is not a problem for NH_3 ($pK_b = 4.75$), which has a lone pair hybridization already very similar to what is needed in NH_4^+.

page 21

(a) NF_3 has smaller angles than NH_3. Why might the angles in PF_3 be *larger* than the angles in PH_3 (page 19)?

Compared to PH_3, PF_3 is larger and has wider angles. Note that this difference in angles is *opposite* that observed in the pair NH_3, NF_3, where NF_3 has the smaller angles. To understand this, realize that the near-90° angles of PH_3 indicate a lone pair with almost pure *s* character already. Changing H to F cannot give it substantially more *s* character and thus cannot make the bonding angles significantly closer to 90°. Instead, the F–F lone pair repulsion spreads their angle.

(b) Why might the angles in PF_3 be smaller than the angles in NF_3 (page 13)?

The smaller angles in PF_3 compared to NF_3 are probably the effect of phosphorus being in the second row and nitrogen being in the first row of the periodic table. As one goes down the periodic table, the splitting between the *atomic* valence *s* and *p* orbitals gets larger and larger. In terms of bonding, this means that it gets harder and harder to involve the *s* orbitals in making bonds to other atoms. Thus, the almost-90° angles in PF_3 indicate that there is virtually no advantage to using the 2*s* orbital on phosphorus for bonding in this molecule.

(c) Compare the structure of PF_3 to that of its oxidation product, POF_3 (page 65). How do you explain the differences?

In POF_3 the P–F distances are smaller, and the F–P–F angle is wider than in PF_3. The net result is that the F–F distances in the two molecules are nearly identical. It's almost as though everything except the central phosphorus atom shifted:

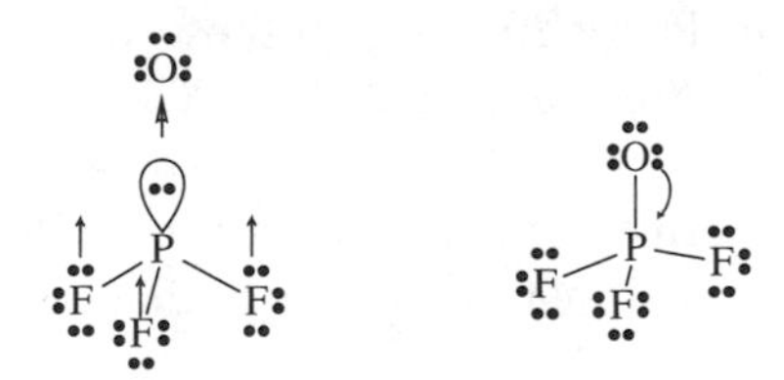

The primary effect of oxidation is to pull electron density away from the central atom, thus making its nucleus effectively stronger. This leads to shorter internuclear distances all around. The wider angle in POF_3 is an indicator that *s* character from the lone pair has been released to the P–F bonding electrons. The very much shorter P–O distance compared to P–F is sometimes taken as a sign that the oxygen atom lone pairs are participating in *pi bonding*.

(d) What is the connection between "oxidation" as discussed here and the sort of oxidation when Fe^{2+} becomes Fe^{3+}?

Oxidation in its most fundamental sense means the removal of electrons. One form of oxidation is by the literal and complete removal of electrons in going from Fe^{2+} to Fe^{3+}. Another, the one we consider here, is more subtle, refering to a slight shift in electron density away from a central atom. Oxidation can also be carried out by reaction with halogen molecules such as F_2 and Cl_2. In that sort of oxidation, electrons initially associated with a central atom (usually a lone pair) are drawn out by the electronegativity of the halogen. See the discussion of the question on page 65 (found on page 209) for more discussion of oxidation and its effect on structure.

(e) What do you predict for the structure of $SiHF_3$?

Both $SiHF_3$ and POF_3 are related to PF_3 by the loss of a strongly *s*-character lone pair. That *s* character goes into bonding, and we might expect shorter distances and wider angles in both POF_3 and $SiHF_3$. But in $SiHF_3$ the central atom also has one less proton, so that would translate into longer bonds. As it turns out, the X–F distances are the same in $SiHF_3$ and PF_3.

(f) Shown on the right is the structure of SF_2, an extremely unstable molecule. Comparing the structures of SF_2 and PF_3 with that of SiF_4 (page 63), what trends in angle and distance do you see for fluorides across the second row? Are these the same trends seen in the series CF_4 (page 33), NF_3 (page 13), OF_2 (page 183)?

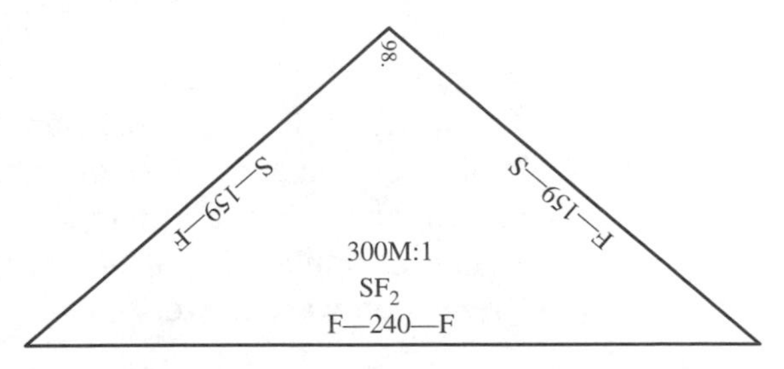

In terms of angle, in comparing the structures of SiF_4 (109.5°), PF_3 (97.8°), and SF_2 (98°), once again we find that there are no clear trends. This is similar to what we found earlier for $CF_4/NF_3/OF_2$ (page 195) and $CCl_4/NCl_3/OCl_2$ (page 196). As for distance, the trend is the same as for the first row fluorides (CF_4, NF_3, OF_2), with distance getting larger, even as we increase the central nuclear charge. Once again, this is likely due to the extreme electronegativity of fluorine.

page 23

(a) Compare the distances in PCl_3 (204 pm) to those in PF_3 (P–F 157 pm) and PBr_3 (P–Br 220 pm). Are the distances in PCl_3 reasonable?

Yes, the distances in PCl_3 are intermediate between those in PF_3 and PBr_3. Cl is below F and above Br on the periodic table, making it intermediate in size.

(b) How does the structure of PCl_3 compare to that of NCl_3 (page 15)?

Just as for PF_3 in comparison with NF_3 (see page 199), PCl_3 is larger than NCl_3 and has smaller angles. Once again we attribute this to phosphorus not being prone to using its 2*s* orbital for bonding unless absolutely necessary.

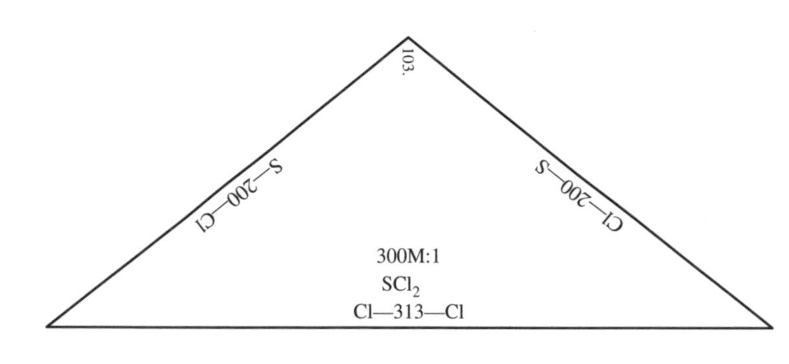

(c) Shown to the left is the structure of SCl_2. What does a comparison of the structures of $SiCl_4$ (Si–Cl 201 pm), PCl_3, and SCl_2 teach us?

There is no trend whatsoever in terms of angle or distance for the series $SiCl_4$, PCl_3, SCl_2. Why should there be? This is consistent with our findings in the case of fluorides. We should be careful about attributing meaning to trends in series of molecules as fundamentally dissimilar as $SiCl_4$, PCl_3, and SCl_2. It is all too easy to overinterpret a limited amount of data.

(d) What would you expect for the structure of $POCl_3$?

Just as in the case of POF_3 in relation to PF_3 (see page 198), $POCl_3$ has smaller P–Cl distances (199 pm) and wider Cl–P–Cl angles (103.5°) than PCl_3. This is the expected effect of oxidation.

(e) What would you expect for the structure of $SiHCl_3$?

In comparing $SiHCl_3$ and PCl_3, we find that the Cl–P–Cl angles in PCl_3 are smaller than the Cl–Si–Cl angles in $SiHCl_3$, and the X–Cl distances in PCl_3 are a little larger. The angle difference makes sense in light of the tendency for a lone pair, especially on phosphorus, to take on extra *s* character, which drives the P–Cl bonds together. The distance difference is similar to that observed for NF_3 and CHF_3, as though here, too, lone pair e^--e^- repulsion were at work.

page 25

(a) How is PHF_2 related to SF_2 in terms of protons and electrons?

PHF_2 has exactly the same number of electrons and protons as SF_2. The only difference is the location of one proton.

(b) Shown to the right is the structure of SF_2. Do the very slight differences in structure of PHF_2 and SF_2 make sense?

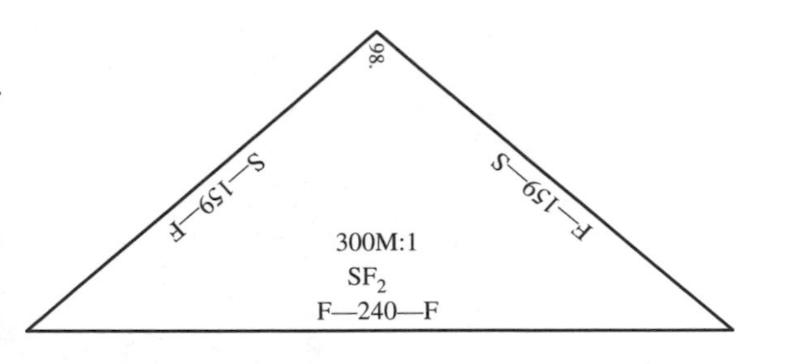

The slightly smaller F–X–F angle in SF_2 indicates that the relocation of the proton into the central nucleus may have slightly reduced the amount of *s* character of the bonding electrons. But the differences in bonding are slight, which may be a clue that it is the *lone-pair electrons* in PHF_2, not the bonding electrons, that are more radically affected by the proton relocation.

(c) Imagine adding an oxygen atom to the phosphorus to make $HPOF_2$. What would you expect its structure to be like?

Oxidation, the pulling of electrons away from the central atom, should result in shorter internuclear distances and more *s*-type bonding character (wider bond angles). Such is the case for the oxidation of PHF_2 to give $HPOF_2$. As it turns out, the F–P–H angle (102° in $HPOF_2$ as compared to 96° in HPF_2) is affected much more than the F–P–F angle, which is 100° in $HPOF_2$ and 99° in HPF_2.

(d) Phosphorus is just below nitrogen on the periodic table. How are the structures of NHF_2 (page 17) and PHF_2 related?

In PHF_2 the distances are all greater than those in NHF_2, and the angles in PHF_2 are all smaller than in NHF_2. This is typical of all second-row/first-row comparisons.

page 27

(a) Why might the structures of IO_3^- and XeO_3 be so similar?

IO_3^- and XeO_3 are *isoelectronic*—they have the same number of electrons.

(b) Why might the distances in IO_3^- be larger than those in XeO_3?

The distances in IO_3^- (180 pm) are larger than those in XeO_3 (176 pm). Note that compared to IO_3^-, XeO_3 has an additional proton, giving its central atom a stronger pull than the one in IO_3^-. It is interesting to compare these systems to IO_4^- (page 75) and XeO_4 (page 77) to see if the effects of oxidation and number of central protons show up there as well.

page 31

(a) Compare the structure of CH_4 to those of NH_4^+ (page 35) and BH_4^- (page 37), both of which also have 10 electrons. How do you explain the differences in size?

BH_4^-, NH_4^+, and CH_4 are isoelectronic. They decrease in size as the number of protons (central pull) increases: BH_4^- is larger than CH_4, which is larger than NH_4^+.

(b) In all of the simple derivatives of methane, the X–C–Y angle never varies by more than just a few degrees from 109.47°. What is so special about 109.47°?

The angle 109.47° is the central angle in a perfect tetrahedron. That is, it is the central angle across the face of a cube:

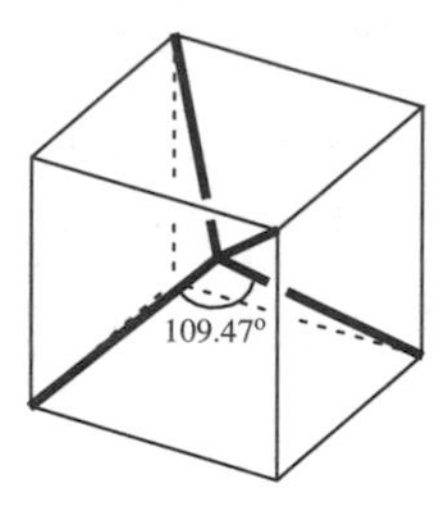

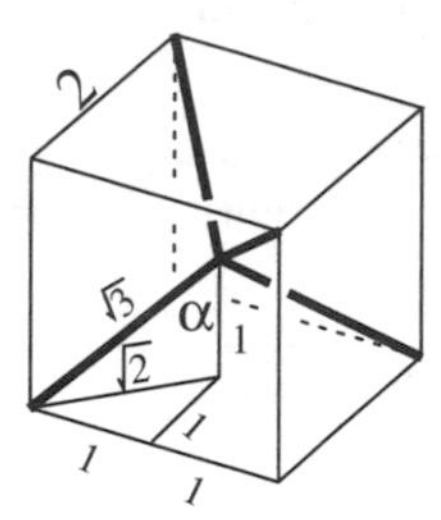

This situation provides the farthest distance four objects can get from each other when tethered to a central point. Another way of expressing this angle is 2α, where $\cos\alpha = 1/\sqrt{3}$.

(c) Methane is the classic tetrahedral molecule with "four equivalent bonds." But are they equivalent? Experiment indicates that they are not, even though all four of the C–H *connections* are identical. How can that be?

This is a loaded question. If you think that just because a molecule is tetrahedral, all of the bonds must be equivalent, then you have fallen into the common trap that awaits those who believe "hybridization" requires localizing of the electrons into four identical spacially directed bonds, each pointing to a different hydrogen atom. My favorite analogy relates the four protons of methane to four people sharing a meal. It goes like this:

> Imagine four people sitting down for lunch. We're interested in knowing what they're having to eat, but, unfortunately, we aren't allowed in the lunch room. Instead, we can only guess what they're eating. As they come out of the room, we ask them what they ate. Each says *exactly the same thing*: "Well, it was sort of like eating an orange—but sort of like eating an apple, too."
>
> One after another they each tell us this same story. Perplexed, we try to imagine what they ate. "Ah, ha!" one of us exclaims, "I know what they had! They've eaten a new *hybrid* fruit—the 'orangapple,' some sort of mix between an apple and an orange. That explains everything!"
>
> Well, that's great, and for a long time we believe that a new fruit actually exists. Some people even tell us that it's reasonable to hybridize fruit in such a way. But no one actually ever gets *evidence* of an orangapple. Then one day we hit on the idea of going in and seeing for ourselves what's left in the room after they come out. What we find is terribly simple and rather embarrassing: three apple cores and the skin of one orange. What they did was to *share* the fruit equally among themselves. Each had three-fourths of an apple and one-fourth of an orange. That's all. No real "hybrid" was necessary. Each had exactly the same lunch.

Similarly, in the case of CH_4, experiments have determined time and again that each C–H *connection* is the same. Nonetheless, the technique of *photoelectron spectroscopy* argues against the existence of four spatially distinct, electronically equivalent hybrid orbitals. Rather, there is evidence for three higher-energy electronic levels in CH_4 and one lower-energy level—exactly what is observed for neon. Hybridization can be simply interpreted as the sharing of atomic orbitals. The four molecular orbitals of methane are constructed from the four valence orbitals of carbon as depicted below:

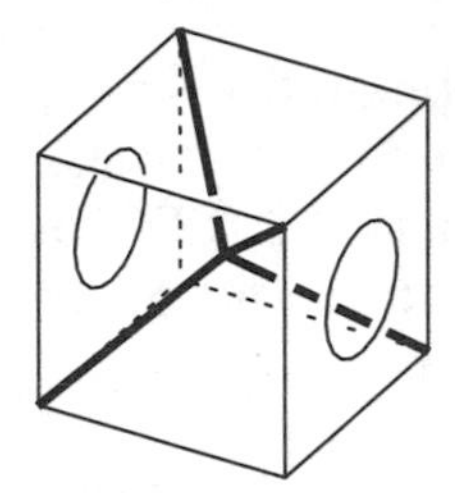
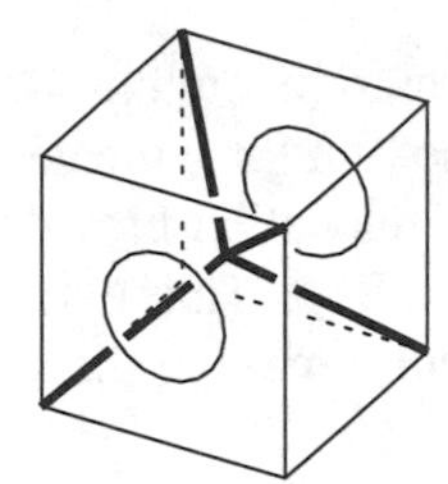
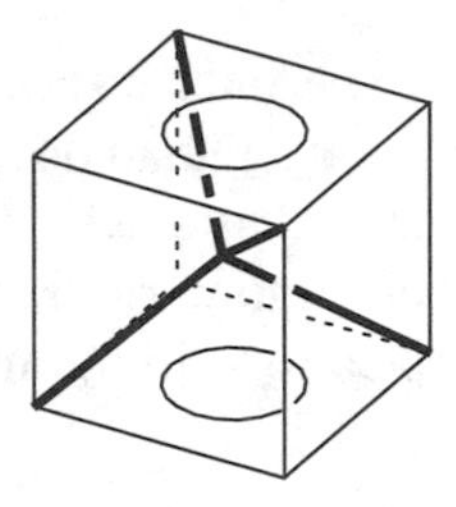

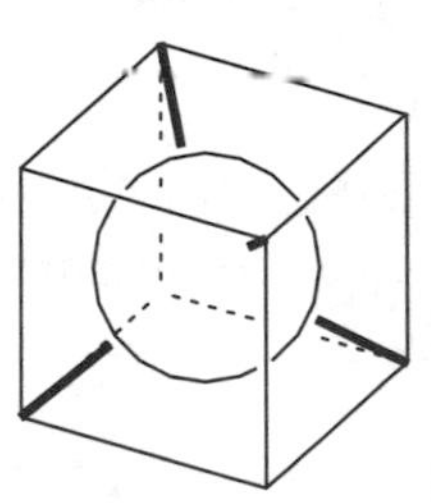

In this view, each *p* orbital is represented by two circles on opposite faces of the cube. Each is lined up directly *between* the lines (not *bonds*!) connecting hydrogens to the central carbon. Each interacts with *all four* of the hydrogens equivalently.

Thus, hybridization does not require localization, the observation of four equivalent protons in CH_4 does not constrain us to four equivalent molecular orbitals, and we are perfectly free to talk about "sp^3 hybridized" orbitals without putting pairs of electrons on the lines of our drawings.

page 33

(a) How does the structure of CF_4 compare to that of CH_4 (page 31)?

CF_4 is exactly the same shape as methane, only larger.

(b) What would you predict for the structures of BF_4^- (page 39) and BeF_4^{-2}?

BF_4^- has one fewer proton than CF_4, so it should be larger. BeF_4^{-2} should be larger still. (It is, with a Be–F distance of 157 pm.)

page 35

The ammonium ion is extremely important in that it is a weak acid, undergoing the following equilibrium quite easily:

$$NH_4^+ + H_2O \rightleftharpoons H_3O^+ + NH_3$$

Which do you think is the weaker acid, H_3O^+ or NH_4^+? That is, which one do you think holds onto its protons more tightly? Why?

Experiments show that NH_4^+ holds onto its protons 10^{11} (one hundred billion) times more strongly than does H_3O^+. That's a lot, and it indicates that NH_4^+ is a much much weaker acid than H_3O^+. The idea of proton relocation provides a possible explanation. H_3O^+ has one more *central* proton than has NH_4^+. This extra nuclear charge attracts all of the electrons in the molecule more tightly, making them less available for bonding.

page 37

(a) How do you explain the observation that BH_4^- is so much larger than CH_4 (page 31)?

Compared to CH_4, BH_4^- has fewer central protons. BH_4^- should be larger than CH_4, which it is.

(b) "H^-" is called *hydride* and consists of two electrons and a proton. BH_4^- is considered a "hydride source" because in many situations it reacts by donating H^- to other molecules:

$$BH_4^- + X \longrightarrow \text{"}BH_3\text{"} + HX^-$$

"BH_3," like "H^-," doesn't actually exist. What is sometimes called BH_3 is really B_2H_6 (page 143). Why might H^- and BH_3 be so reactive as to not really exist?

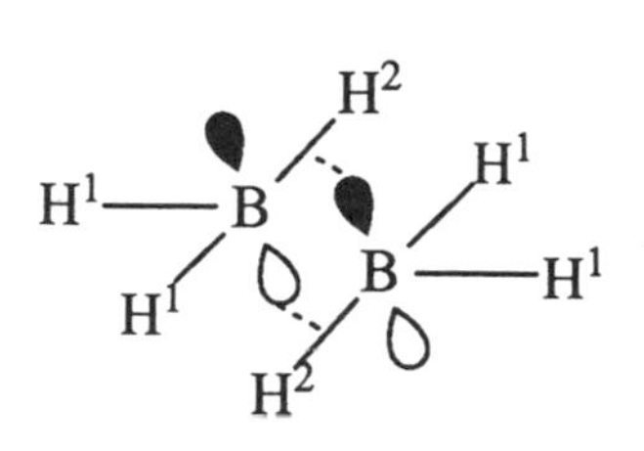

The two electrons in H^-, held together by just one proton, are in great need of stabilization. Just about any molecule will do, as long as it has an available empty orbital. BH_3 is just such a hypothetical molecule, with its vacant central *p* orbital. Thus, we find BH_4^- is stable relative to isolated BH_3 and H^-. That vacant *p* orbital in BH_3 is particularly available, because the hydrogen atoms can't take advantage of it (as might be the case in BF_3, for example.) Thus, BH_3 even reacts with a copy of itself to form B_2H_6, allowing some of the electrons in each molecule to be stabilized. On the model of B_2H_6 (page 143), you can see that the outer H–B–H angle is still around 120°—in fact, it's a little *larger* than 120°.

page 39

(a) Why is BF_4^- so much larger than BH_4^- (page 37)?

All X–F bonds are longer than the corresponding X–H bonds. Fluorine is using its 2*p* orbital for bonding, whereas hydrogen is using its 1*s* orbital.

(b) Why is BF_4^- so much smaller than BeF_4^{-2} (Be–F 157 pm)?

With fewer central protons, BeF_4^{2-} should be larger still than BF_4^-. Indeed, the Be–F distance in BeF_4^{2-} is a whopping 157 pm, making it larger even than SiF_4.

(c) Compare the structure of BF_4^- to that of BF_3 (see page 13). How do you explain the differences?

The B–F distances in BF_4^- (141 pm) are much larger than the B–F distances in BF_3 (130 pm). There are three possible reasons for this difference, all of which probably contribute to the real situation. First, simple electrostatics: The four fluorine atoms in BF_4^- all share the negative charge and repel each other more than the three fluorines in BF_3. Second, rehybridization: The *s* character of the boron atomic orbitals in BF_3 is being divvied up among three fluorines, whereas in BF_4^- it is shared by four. That means each B–F connection gets less *s* character in BF_4^-, and less *s* character generally means longer connections. Finally, delocalization: The empty *p* orbital in BF_3 is not really empty. Rather, calculations suggest that the "lone pairs" on each fluorine get to share this space, and thus get a little extra stabilization. That sharing, called a *delocalized pi bond*, draws together the atoms in BF_3 and is not possible in BF_4^-, since there is no empty *p* orbital to share.

page 41

(a) *Lewis bases* donate pairs of electrons when reacting; *Lewis acids* accept them. This complex is the classic example of the result of a Lewis acid/base interaction. What was the Lewis acid and what was the Lewis base?

In forming the complex $BF_3 \cdot NH_3$, the Lewis base is NH_3; the Lewis acid is BF_3.

(b) In this study the N–H distances were not precisely determined. Would you expect them to be larger or smaller than the N–H distances in NH_3?

Since the NH_3 part of this complex is donating electrons, it must end up with a slight positive charge. Almost certainly this results in the same sort of situation as occurs in NH_4^+, which has larger N–H distances than has NH_3.

(c) In comparison to BF_3 (see page 13), the B–F distances in $BF_3 \cdot NH_3$ are significantly longer. Why?

Just as for BF_4^- in relation to BF_3 (see above), the answer is threefold, involving electrostatic repulsion of fluorine atoms, rehybridization from "sp^2" to "sp^3" bonding, and loss of pi-electron delocalization.

page 43

(a) Although BH_3 does not exist by itself, here is an example of one of its stable complexes. Can you write an equation making this complex from B_2H_6 and PF_3?

Here it is: $B_2H_6 + 2\,PF_3 \longrightarrow 2BH_3 \cdot PF_3$

(b) In the analogous structure, CH_3–SiF_3, the C–Si distance is known to be 188 pm. How do you explain the smaller distance of 184 pm for B–P in $BH_3 \cdot PF_3$?

Although perhaps at first sight CH_3–SiF_3 and $BH_3 \cdot PF_3$ have little in common, they differ only in the placement of one proton. The B–P bond is more *ionic* than the C–Si bond, giving the connection between boron and phosphorus an extra-strong component. That translates in this case and in many, many others into a smaller interatomic distance.

page 45

Compare the structure of $BH_3 \cdot PF_3$ to that of POF_3 (page 65). Both of these molecules can be thought of as PF_3 molecules that have reacted with electron-deficient "units" (O or BH_3). How are they similar in respect to PF_3?

The effect of borane complexation on PF_3 is similar to the effect observed for oxidation of PF_3 to POF_3. In both cases the lone pair on phosphorus gets involved in bonding. In fact, $PF_3 \cdot BH_3$ and PF_3O both have the same number of protons and electrons! Not surprisingly, relocating three hydrogens into the boron nucleus to give oxygen results in a much stronger attraction for the bonding electrons in POF_3. The P–O distance in POF_3 (144 pm) is much, much smaller than the P–B distance in the borane complex (184 pm).

page 47

(a) Would you expect the B–H distances in the hypothetical BH_3 molecule to be larger or smaller than the B–H distances in $BH_3 \cdot CO$?

In forming this complex, CO was the Lewis base and BH_3 was the Lewis acid. This is similar to the formation of BH_4^- from BH_3. (See the discussion on page 204.) As such, the boron in $BH_3 \cdot CO$ is somewhere between BH_3 and BH_4^- in character. Since the B–H connections in $BH_3 \cdot CO$ (119 pm) are smaller than those in BH_4^- (125 pm), it follows that the B–H connections in BH_3 should be smaller still.

(b) The C–O distance in $BH_3 \cdot CO$ is 113 pm. This is exactly the distance in carbon monoxide itself. Why might this be?

The principal orbital involved when CO acts as a Lewis base is the lone pair situated primarily on the carbon. From a molecular orbital point of view, this lone pair is really delocalized over the molecule. It is the "highest occupied σ molecular orbital" (σHOMO) and has largely *nonbonding* character.

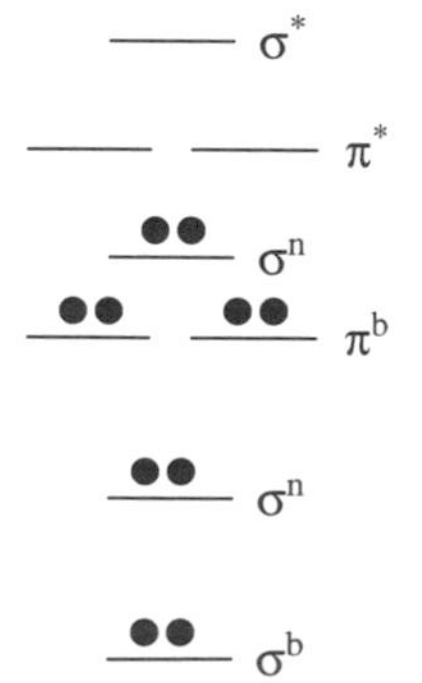

Most introductory chemistry books assign this orbital as σ^b, presumably to make life easier, but that is a mistake based on forgetting to mix *s* and *p* atomic orbitals. A corrected version of the molecular orbital diagram for 10-e^- diatomic molecules is shown here. Note the inclusion of two orbitals labeled σ^n, which means "sigma nonbonding." In CO, both of these orbitals are filled. These are the lone pairs; the lower-energy pair being weighted toward the oxygen atom, and the higher-energy pair being weighted toward carbon. Whether the highest σ orbital is slightly bonding or slightly antibonding and whether it resides slightly above or below the π^b orbitals in energy depends upon subtle characteristics of the bonded atoms and is of minimal practical significance.

Two forms of evidence point to the nonbonding character of the σHOMO in diatomic species with 10 valence electrons such as CO, CN^-, and N_2. First of all, photoelectron spectroscopy, in which the energy it takes to remove an electron from a molecule is recorded, indicates that in these molecules very little change in bond strength occurs when the highest σ-type electron is removed. Second, the bond distances are known not only for CO, N_2, and CN^-, but also for CO^+, N_2^+, and CN, where the tenth electron is missing. Very little change occurs in removing that last electron:

10-e^- Species	9-e^- Species	Distance Change	% Change	σHOMO Character
CO 113 pm	CO^+ 112 pm	slightly shorter	−1%	slightly antibonding
N_2 110 pm	N_2^+ 112 pm	slightly longer	+2%	slightly bonding
CN^- 115 pm	CN 117 pm	slightly longer	+2%	slightly bonding

Thus, the highest-energy electrons in CO, N_2, and CN^- are best described just exactly as one would using Lewis structures: as lone-pair electrons. It's just that in molecular orbital terms, these "lone pairs" are delocalized over both atoms, exactly as in the case of the outer oxygen lone pairs of CO_2, HNO_2, and H_2CO_3.

It is interesting to compare these data to species where the electron being removed is coming from a truly antibonding orbital:

No. of e^-s	n-Electron Species	(n–1)-Electron Species	Distance Change	% Change	σHOMO Character
11	NO 115 pm	NO^+ 106 pm	much shorter	8%	strongly antibonding
12	O_2 121 pm	O_2^+ 112 pm	much shorter	8%	strongly antibonding
13	O_2^- 128 pm	O_2 121 pm	much shorter	6%	strongly antibonding
14	O_2^{-2} 149 pm	O_2^- 128 pm	much shorter	13%	strongly antibonding

In these cases, removal of an electron actually *shortens* the bond distance by as much as 13%. *This* is what one expects from an antibonding electron!

page 49

> The alternative conformation for ethane is called *eclipsed*, in which the front CH_3 in the Newman projection is turned 60°. In the eclipsed conformation the three front H atoms line up with the three H atoms in the back. Why might the staggered conformation be more stable than the eclipsed conformation?

The difference in energy between the two forms of ethane is very slight—only a few kcal/mol—and must be attributable to very subtle factors. Suffice it to say here, the only really adequate explanation for this difference in energy is provided by a *delocalized* picture of ethane, in which all electrons are distributed into orbitals that are spread out all through the molecule. For a thorough discussion, see the article by Nobel laureate Roald Hoffmann in *The Journal of Chemical Physics*, volume 39, page 1397 (1963).

page 53

(a) If you look carefully at the model of CH_3NH_2, you will notice that the H–C–N angles are not all the same. The one to H^a is 113°, while the ones to H^b are only 107.9°. That is, the NH_2 group leans away from H^a. What is going on here?

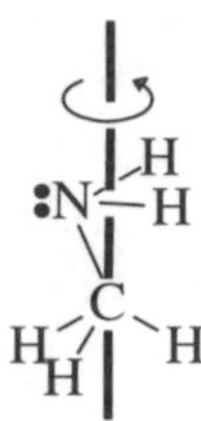

The NH_2 group of CH_3NH_2 was found by microwave spectroscopy to rotate *asymmetrically*, as shown to the left. That is, it precesses (wobbles), with the nitrogen atom on one side of the axis and the two hydrogen atoms on the other. This precession results in a slightly wider angle to one of the hydrogen atoms of the CH_3 group.

(b) Which is flatter around the nitrogen atom, CH_3NH_2 (see page 17) or NH_3 (page 11)? What does this mean for the lone pair?

Compared to NH_3, CH_3NH_2 is flatter around the nitrogen, indicating a higher-energy (more *p*-character) lone pair. Mostly, this is the effect of CH_3 being a larger group and thus requiring more space than a hydrogen atom.

(c) Which would you predict to be more basic, CH_3NH_2 or NH_3?

CH_3NH_2, with a pK_b of 3.3, is somewhat more basic than NH_3 ($pK_b = 4.75$).

page 55

(a) Once again, careful examination of the model indicates that the OH leans away from one of the hydrogen atoms. Why might this be the case?

Again, as for the NH_2 group in CH_3NH_2, the OH group in methanol wobbles as it rotates.

(b) Why is the C–O distance in CH_3OH smaller than the C–N distance in CH_3NH_2 (page 53)?

Once again, the only difference between CH_3OH and CH_3NH_2 is the location of a single proton. If you consider the "central atom" to be oxygen in CH_3OH and nitrogen in CH_3NH_2, then just as for H_2O and NH_3, the distances in the molecule with oxygen are shorter. Similarly, the C–O distance in CH_3OH (142 pm) is longer than the C–F distance in CH_3F (139 pm).

page 57

How much difference would you expect between the structure of CHF_3 and that of $SiHF_3$?

$SiHF_3$ is much larger than CHF_3, but the two molecules are virtually identical in all central angles. The F–Si–F angle in $SiHF_3$ (108.3°) is insignificantly tighter than the F–C–F angle in CHF_3 (108.8°).

page 59

(a) Chloroform is weakly acidic ($pK_a \approx 25$) and reacts with strong bases such as NaOH to form CCl_3^-. How are NCl_3 and CCl_3^- similar?

NCl_3 and CCl_3^- are isoelectronic; NCl_3 has an extra proton.

(b) What is the probable structure (shape, distances, angles) of CCl_3^-?

The X–Cl distances in CCl_3^- should be larger even than those in $CHCl_3$ (177 pm), since the bonding electrons in CCl_3^- are slightly better shielded without the H proton. Its Cl–C–Cl angles should be tighter than those in $CHCl_3$ (112°), once again (as with NCl_3) because the lone pair will want more than its fair share of the central atom's *s* character. Recent calculations estimate the C–Cl distance in CCl_3^- to be 187 pm and the Cl–C–Cl angle in CCl_3^- to be 104° [Gutsev, G. L., *J. Chem. Phys.* **95**, 5773 (1991)].

page 63

> Imagine that one of the protons in one of the F atoms of SiF_4 were somehow moved over to join with the Si atom. What molecule would you now have? What do you predict for its structure? (See page 65.)

Relocating one of the F protons into Si changes an Si–F connection to a P–O connection and SiF_4 to POF_3. The stronger central nuclear pull makes for much shorter connections all around, especially for the oxygen (144 pm for P–O and 152 pm for P–F in POF_3 vs. 154 pm for Si–F in SiF_4). The remaining P–O connection takes more than its share of the *s* character of bonding. The fluorines are left with more of the *p* character and, consequently, the F–P–F angle in POF_3 (101°) is smaller than the F–Si–F angle in SiF_4 (109°).

page 65

> It is interesting to compare the structures of PF_3 and POF_3. What other molecules and ions in this book are similarly related by oxidation? Do they show the same trends?

There are many such examples. The basic trend seems to be that the addition of an oxygen atom shortens the bond distances and widens the angles. Several other species that are related by oxidation can be found in this book. Here is the whole list. You decide if the trend is "real" or not.

Reduced Form	A–X (pm)	X–A–X	Oxidized Form	A–X (pm)	X–A–X
PF_3 (page 21)	$157_{P–F}$	97.8	POF_3 (page 65)	$152_{P–F}$	$101.3_{F–P–F}$
XeO_3 (page 27)	176	$103._{O–Xe–O}$	XeO_4 (page 77)	174	(109.5)
SF_4 (page 83)	$164_{S–Fax}$	$89._{Feq–S–Fax}$	SOF_4 (page 91)	$158_{S–Fax}$	$89.6_{Feq–S–Fax}$
	$154_{S–Feq}$	$103._{Feq–S–Feq}$		$155_{S–Feq}$	$110._{Feq–S–Feq}$
CO (page 177)	113	—	CO_2 (page 179)	116	(180.)
NOF (page 181)	$152_{N–F}$	$110._{F–N–O}$	NO_2F (page 187)	$147_{N–F}$	$112._{F–N–O}$
NO_2^- (page 181)	124	115.4	NO_3^- (page 187)	124	(120.)
SO_2 (page 181)	143	119.5	SO_3 (page 187)	143	(120.)
XeF_4 (page 189)	195	(90.)	$XeOF_4$ (page 97)	190	$89.9_{F–Xe–F}$

Note that in SF_4 there are two kinds of S–F distances. Only one shows the trend. Other exceptions are CO/CO_2 and compounds having more than one lone pair before oxidation (for example, SF_2/SOF_2 or $HPF_2/HPOF_2$, which are not shown in the table).

page 67

(a) The structure of H_3PO_4 given here was based on x-ray analysis of a crystal. If you look carefully, you will see that there are three *different* OH groups, each with slightly different angles to the P–O bond. What could cause this variation?

Molecules in crystals are exposed to forces from *crystal packing*—that is, the three-dimensional arrangement of molecules within the crystal. These forces lead to permanent distortions which are only transiently found in gas-phase molecules.

(b) Compare this structure to that of H_2SO_4 (page 71). Why is the S–O distance smaller than the P–O distance?

The P–O distance (151 pm) and P–OH distance (156 pm) in H_3PO_4 are considerably longer than the corresponding distances in H_2SO_4 (S–O 143 pm and S–OH 154 pm). That extra central proton in H_2SO_4 makes a *big* difference.

page 69

(a) This is a slightly different structure from the previous one, also found in the same crystal. What lesson can we learn from this comparison?

Looking at variations in distances and angles among different crystal structures, or even among different molecules and ions in the same crystal, helps us to recognize the effects of crystal packing forces. Bond angles are especially "soft"; angles in gas-phase samples can be significantly different from angles in crystalline samples.

(b) Why would the P–O distance in H_3PO_4 be larger than the P–O distance in POF_3 (page 65)?

Compared to POF_3, the P–O distance in H_3PO_4 is considerably larger (151 pm in H_3PO_4 vs. 144 pm in POF_3). Part of this huge difference is due to the fact that the H_3PO_4 is in a crystal, whereas in the POF_3 study the molecules were in the gas phase. The lone oxygen of H_3PO_4 in the crystal is interacting with two hydrogens on other H_3PO_4 molecules. Almost certainly, though, there is more to it than that. Fluorine is highly electronegative, leading to a more electron-deficient central atom in POF_3 than in H_3PO_4. This electron deficiency gives the central atom a stronger pull on all other bonding electrons, leading to a shorter P–O distance in POF_3. In fact, fluorine is so electronegative, that the phosphorus center in POF_3 is almost as positive as the sulfur in H_2SO_4, which has an S–O distance of 143 pm.

page 71

Why might the O–S–O angle between non-OH oxygens in H_2SO_4 be so large compared to any of the other angles?

The large O–S–O angle in H_2SO_4 (119°) is very similar to that found in SO_2F_2, SO_2Cl_2, and ClO_3F, all of which are near 120°. Remember, oxidation is the removal of electrons from the central atom. That means that the oxygen atoms in all of these compounds all bear extra electron density. The simple explanation is that this extra electron density leads to extra electron repulsion between oxygen atoms, and thus to wider angles. Also working here, though, is the effect of pi bonding to the central atom by the oxygen lone pairs. Once again, we are probably at the limits of legitimacy for our theoretical model and should be cautious.

page 73

> In ClO_3^-, the Cl–O distances are 157 pm, and the O–Cl–O angle is 106.7°. Is this what you would expect?

The smaller distances and wider angles in ClO_4^- as compared to ClO_3^- are consistent with what we have seen for oxidation.

page 75

> Why aren't the angles in IO_4^- all 109.47°?

The angles in IO_4^- are not all 109.47° because of distortions involving the crystal lattice. This distortion involves "squashing" the tetrahedron along an axis bisecting two opposite O–I–O angles and are attributable to Na^+ ions in the vacinity of the oxygen atoms.

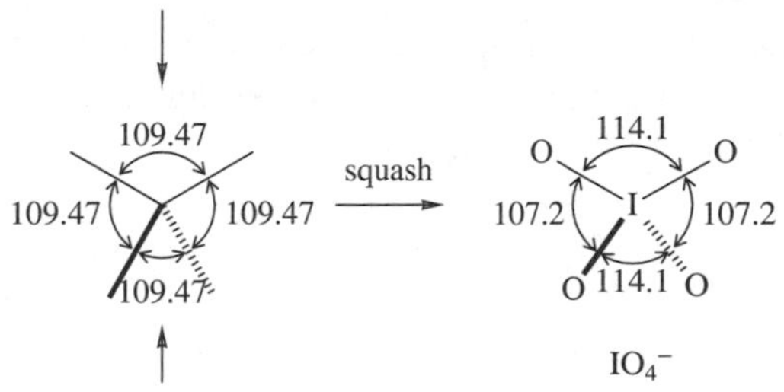

page 77

> How are IO_4^- and XeO_4 related? Is 174 pm reasonable for the Xe–O distance in XeO_4?

XeO_4 has an extra proton in the center as compared with IO_4^-, but otherwise they are the same (discounting neutrons). As a result, the Xe–O distance in XeO_4 (174 pm) is slightly smaller than the I–O distance in IO_4^- (178 pm).

Summary

Several trends in structure are evident from these data, most of which have been brought out in one way or another as part of this discussion. How many did you find? A few are only evident when you combine these data with data for more complex molecules (Parts 2–4) and data for one- and two-dimensional molecules (Part 6). Here's my list:

Trend 1 — For two species differing only by the row of the periodic table for the **central atom**, the one having the central atom, A, further down on the periodic table has longer A–X bonds and tighter X–A–X angles.

Examples include NH_3 vs. PH_3, NF_3 vs. PF_3, and NCl_3 vs. PCl_3.

Trend 2 — For two species differing only by the row of the periodic table for an **outer atom**, the one having the outer atom, X, further down on the periodic table has longer A–X bonds and wider X–A–X angles involving that atom.

Examples include NF_3 vs. NCl_3, PF_3 vs. PCl_3, and NOF vs. NOCl.

Trend 3 — For two species differing only by the number of protons in the central nucleus, A, the one having more central protons has shorter A–X bonds.

Examples include IO_3^- vs. XeO_3, BH_4^- vs. CH_4, CH_4 vs. NH_4^+, CO vs. NO^+, and CO_2 vs. NO_2^+.

Trend 4 — For two species differing only by the number of protons in an outer nucleus, X, the one having more outer-nucleus protons has a longer A–X bond to that atom, shorter A–X′ bonds to other atoms, and tighter X–A–X′ angles.

Examples include N_2O vs. NO_2^+, NO_2^- vs. NOF, and NO_3^- vs. NO_2F.

Trend 5 — For two species differing only by the location of a proton—attached to the central atom as ${}_{n}$A–H or in the central nucleus as ${}_{n+1}$A:—the species with the proton in the nucleus (and thus having an additional central-atom lone pair) has tighter X–A–X angles. A–X distances in that species will depend upon the attached atoms. If the attached atoms are hydrogens, distances will be shorter; if the attachments are F or OH, distances will be longer.

This is really a very little recognized trend, especially for X=F. For X=H, examples include CH_4 vs. NH_3, NH_3 vs. H_2O, H_2O vs. HF, SiH_4 vs. PH_3, and PH_3 vs. H_2S. For X=F, examples include NHF_2 vs. OF_2, PHF_2 vs. SF_2, and CHF_3 vs. NF_3.

Trend 6 — For two species differing only by the location of a proton—attached to an outer atom as ${}_{n}$X–H or in that outer atom's nucleus as ${}_{n+1}$X:—the species with the proton in the nucleus (and thus having an additional outer-atom lone pair) has shorter A–X and A–X′ bonds and tighter X–A–X′ angles.

Examples include $PF_3{\cdot}BH_3$ vs. POF_3, H_3PO_4 vs. POF_3, C_2H_2 vs. HCN, HNCO vs. CO_2, and HN_3 vs. N_2O. Several exceptions exist, primarily resulting from further involvement of the lone pair. For example, in comparing allene, $CH_2{=}C{=}CH_2$, with ketene, $CH_2{=}C{=}O$, the C=O bond in ketene is shorter than the C=C bond in allene. However, the C=C bond in ketene is *longer* than the C=C bond in allene.

Trend 7 For two species differing only by the location of a proton—in an outer-atom nucleus as ${}_{n}A{-}{}_{m}X$ or in the central nucleus as ${}_{n+1}A{-}{}_{m-1}X$—the species with the proton in the central nucleus has the shorter A–X bond.

Examples include SiF_4 vs. POF_3, PF_5 vs. SOF_4, CO vs. N_2, CO_2 vs. N_2O, BF_3 vs. CF_2O, and CF_2O vs. NO_2F. One exception is CO^+ vs. N_2^+, where the very short C–O distance (112 pm) is no larger than the N–N distance in N_2^+.

Trend 8 Oxidation of the central atom by the replacement of a lone pair in $:AX_n$ with a bond to an oxygen atom, giving OAX_n, generally leads to shorter A–X bonds and wider X–A–X angles.

Examples of this trend include PF_3 vs. POF_3, XeO_3 vs. XeO_4, and NOF vs. NO_2F. There are lots of exceptions to this trend. For more examples, see the table on page 209.

Trend 9 Oxidation of the central atom by the replacement of a lone pair in $:AX_n$ with bonds to two fluorine atoms, giving F_2AX_n, leads to shorter A–X distances.

Examples include PF_3 vs. PF_5, SF_2 vs. SF_4, SF_4 vs. SF_6, and BrF_3 vs. BrF_5. In cases where more than one type of X–F bond exists initially, sometimes one gets smaller while another actually gets longer. Several examples are shown below. In each case, the bonds that get longer seem to be those *opposite* to the reactive lone pair.

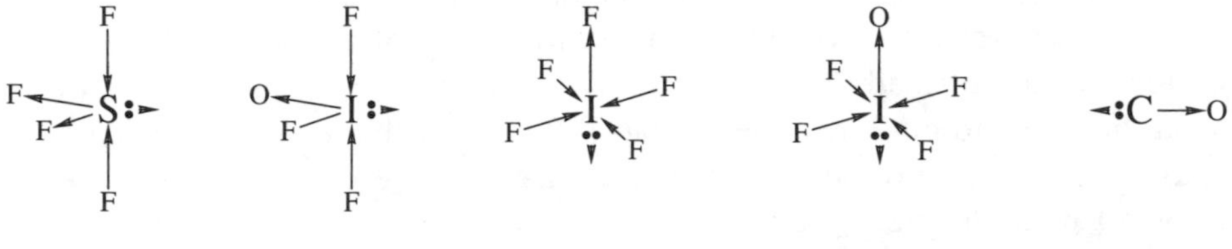

Ten recurrent ideas seem to be sufficient for rationalizing these trends. They are summarized below:

(1) Bonding is primarily an *electronic* phenomenon.

(2) Electrons derive stability by being associated with protons; for molecules such as CH_4, protons in the *central nucleus* are more effective in stabilizing electrons than protons in the outer hydrogen atoms.

(3) Some electrons (the *core* electrons) are so stabilized by a single nucleus that they are not significantly involved in bonding, while others, the *valence electrons*, benefit from association with more than one atom. Among these valence electrons are *bonding* and *nonbonding* electrons.

(4) Nonbonding valence electrons that primarily reside on one atom (lone pairs) are very important in determining the shape and reactivity of molecules.

(5) Valence electrons in most molecules and ions can be thought of as being paired in *molecular orbitals*, which can be thought of as "existing" independently of whether electrons are in them. This leads to the idea of *unoccupied molecular orbitals*, which is also helpful in understanding much of molecular structure and reactivity.

(6) We generally think of molecular orbitals as being constructed from *atomic orbitals*, which are themselves analogues of the orbitals known exactly only for the hydrogen atom. This construction is highly approximate, based solely on the need for manageable concepts and timely computer calculations.

(7) Molecular shapes, especially *angles*, are clues to the *s* and *p* character of the orbitals used for bonding—wider bond angles being associated with more *s* character.

(8) Atomic and molecular orbitals are characterized not only by shape, but also by energy. Some atomic orbitals (the *s* orbitals) are lower in energy than others (the *p* orbitals), and this energy difference is also reflected in the molecular orbitals arising from them. Thus, molecular orbitals having more *s* character are lower-energy molecular orbitals.

(9) The most meaningful comparisons can be made among systems having the same number of electrons and protons, or at least the same number of valence electrons.

(10) Everything that we know or believe about molecules is an approximation based on imperfect models and incomplete understanding.

More than anything, the lesson here should be a caution about trying to explain all of the nuances of molecular structure using such simplistic ideas as proton relocation, "VSEPR" theory, or even simple molecular orbital theory. It is quite possible that the question, "Why?" will never be answered for the vast majority of cases of molecular structure. In the "real world" of studying molecular structure, chemists have taken a different tack. Instead of trying to pin down the various influences behind what we observe, we simply say that the driving force behind *all* molecular structure is the natural tendency to reach the lowest possible energy state.

Perhaps someday we will look back on *s* and *p* and *d* orbitals and say, "My, how naive." But for now they are the basis of much of our understanding. The key is to try to keep all of our explanations in perspective and to realize that as soon as we even utter the word "bond" or "orbital" we are putting molecules into a box that is of our making, not nature's.

Sources and Methods

References for all data in Parts 1–6 can be found here. References are given in the following format:

Journal/Book.Year.Volume.Page.Method

Journal names are abbreviated according to the following key:

AC	Acta Crystallographica
ACS	Acta Chemica Scandinavica
AJ	Astrophysics Journal
AM	American Mineralogist
CJP	Canadian Journal of Physics
CJR	Canadian Journal of Research
DFS	Discussions of the Faraday Society
IC	Inorganic Chemistry
JACS	Journal of the American Chemical Society
JCP	Journal of Chemical Physics
JCS	Journal of the Chemical Society
JCSJ	Journal of the Chemical Society of Japan
JFIC	Journal of Fluorine Chemistry
JMS	Journal of Molecular Spectroscopy
JMSt	Journal of Molecular Structure
JOMC	Journal of Organometallic Chemistry
JOSA	Journal of the Optical Society of America
JPC	Journal of Physcial Chemistry
JSSC	Journal of Solid State Chemistry
PR	Physical Review
PIAS	Proceedings of the Indian Academy of Science
RTC	Recueil des Travaux chimiques des Pays-Bas et de la Belgique
SA	Spectrochimica Acta
Sc	Science
SAWW	Sitzungsber.Akad.Wiss.Wien
TFS	Transactions of the Faraday Society
TKB	Tidsskrift for Kjemi, Bergvesen of Metallurgi
ZaC	Zeitschrift für Anorganische und Allgemeine Chemie
ZK	Zeitschrift für Kristallographie
ZPC	Zeitchrift für Physikalische Chemie

Several books that include compilations of data were also helpful in finding many of the primary journal references. They are listed below. If a reference for a specific molecule or ion is from one of these books, the first author's name is used instead of a journal name.

Bowen, H. J. M.; Donohue, J.; Jenkin, D. G.; Kennard, O.; Wheatley, P. J.; Whiffen, D. H. *Table of Interatomic Distances and Configuration in Molecules and Ions*; Chemical Society: London, 1958.

Hehre, W.; Radom, L.; Schleyer, P.; Pople, J. *Ab Initio Molecular Oribital Theory*; John Wiley & Sons: New York, 1986.

Herzberg, G. *Molecular Spectra and Molecular Structure: I. Infrared Spectra of Diatomic Molecules*, Second Edition; Van Nostrand: New York, 1950.

Herzberg, G. *Molecular Spectra and Molecular Structure: II. Infrared and Raman Spectra of Polyatomic Molecules*; Van Nostrand: New York, 1945.

Herzberg, G. *Molecular Spectra and Molecular Structure: III. Electronic Spectra and Electronic Structure of Polyatomic Molecules*; Van Nostrand: New York, 1966.

Hueckel, W. *Structural Chemistry of Inorganic Compounds, Volume II*; Elsevier: Amsterdam, 1951.

Peckett, A. *The Colours of Opaque Minerals*; Van Nostrand Reinhold: New York, 1992.

Trotman-Dickenson, A. F. *Comprehensive Inorganic Chemistry, Volumes 1–5*; Pergamon Press: Oxford, 1973.

Wells, A. F. *Structural Inorganic Chemistry,* Fifth Edition; Oxford University Press: London, 1984.

Wyckoff, R. *Crystal Structures, Volume I*, Second Edition; Interscience: New York, 1965.

The methods used to determine the structure of the specific molecules and ions are indicated by a single letter following the reference and briefly described below:

E **Electron Diffraction** High-energy electrons of wavelength around 6 pm are diffracted by the nuclei of gaseous molecules (and, more recently, specially prepared thin solid layers) The intensity data are transformed to give a "radial distribution curve," which can be analyzed in terms of molecular interatomic distances once atom-based effects are subtracted.

I **Infrared Spectroscopy** Primarily involving small molecules, infrared absorption is at an energy which excites molecular vibrations. Fine structure also gives information about rotational states of the molecule and can be interpreted in terms of bond distances and bond "force constants."

M **Microwave Spectroscopy** Microwaves are lower in energy than infrared radiation and directly excite molecular rotational states. Comparing the absorptions of molecules that are isotopically enriched (and thus slightly different in mass) allows very precise measurements to be made, which can then be fitted to sets of parameters including interatomic distances and angles.

N **Neutron Diffraction** High-energy neutrons of wavelength approximately 140 pm are diffracted by the nuclei of solids and liquids. Analysis of the resultant diffraction intensities reveals structural information, especially the positions of atoms of low mass such as hydrogen, which are often difficult to locate by x-ray diffraction.

Q **Nuclear Magnetic Resonance Spectroscopy** The structures of two ions, NH_4^+ and BH_4^-, were determined by analysis of single crystals by ^1H NMR spectroscopy at –195 °C (the temperature of liquid nitrogen). At this temperature vibrations are frozen out and the lineshape of the resonance can be used to give the distance between hydrogen atoms in the sample.

R **Raman Spectroscopy** In this method, monochromatic visible light from a laser is passed through a gaseous sample. This beam is not absorbed by the sample, but does induce a dipole through polarization. The interaction of the visible light with the sample leads to scattering. The scattered light frequencies and intensities are measured and are found to be related to the vibrational and rotational characteristics of the molecule or ion under investigation.

UV **Ultraviolet Spectroscopy** For simple diatomic or linear molecules and ions, ultraviolet absorptions or emissions have fine structure that can be interpreted in terms of vibrational and rotational states. In other words, this method is another way of getting the same information that comes from microwave and infrared spectroscopy.

X **X-ray Diffraction** In this method, used primarily for probing the structure of crystalline solids, x-rays ($\lambda \approx 100$ pm) are diffracted primarily by the electrons in the solid—especially core electrons. The intensities of the diffracted beam at different angles are recorded and related to structure by an iterative analysis.

Page	Formula	Reference
11	NH_3	ACS.1955.9.815.E
13	NF_3	JACS.1950.72.1182.E
15	NCl_3	ZaC.1975.413.61.X
17	NH_2CH_3	JCP.1955.23.1735.M
17	NHF_2	JCP.1963.38.456.M
19	PH_3	JCP.1959.31.449.M
21	PF_3	IC.1969.8.867.E
23	PCl_3	JCP.1950.18.1109.M
25	PHF_2	JACS.1968.90.1705.M
27	IO_3^-	JCP.1971.54.2556.X
27	XeO_3	JACS.1963.85.817.X
31	CH_4	CJP.1955.33.138.R
33	CF_4	JCP.1953.21.565.E
35	NH_4^+	JCP.1954.22.643.Q
37	BH_4^-	DFS.1955.19.230.Q
39	BF_4^-	AC.1971.B27.1102.X
41	$BF_3 \cdot NH_3$	AC.1951.4.369.X
43	$BH_3 \cdot PF_3$	JCP.1967.46.357.M
47	BH_3CO	PR.1950.78.512.M
49	CH_3CH_3	ACS.1955.9.815.E
53	CH_3NH_2	JCP.1955.23.1735.M
55	CH_3OH	JCP.1955.23.1739.M
57	CHF_3	JCP.1952.20.605.M
59	$CHCl_3$	JCSJ.1946.67.93.E
61	SiH_4	JCP.1955.23.922.I
63	SiF_4	JACS.1934.56.2373.E
65	POF_3	IC.1971.10.344.E
67	H_3PO_4	AC.1974.B30.1470.X
71	H_2SO_4	AC.1965.18.827.X
73	ClO_4^-	PIAS.1957.A56.134.X
75	IO_4^-	AC.1970.B26.1782.X
77	XeO_4	JCP.1970.52.812.E
83	SF_4	JCP.1963.39.3172.M
85	SeF_4	JMS.1968.28.454.M
89	PF_5	IC.1965.4.1775.E
91	SOF_4	JCP.1969.51.2500.M

Page	Formula	Reference
95	BrF_5	JCS(CC).1971.–.1567.M
97	$XeOF_4$	JMS.1968.26.410.X
101	SF_6	ZPC.1933.B,21.297.E
103	PF_6^-	AC.1956.9.825.X
111	UF_7^{-3}	AC.1954.7.783.X
115	NbF_7^{-2}	AC.1966.20.220.X
119	$UO_2(NO_3)_3^-$	AC.1965.19.205.X
123	ZrF_8^{-4}	JCP.1964.41.3478.X
127	ZrF_8^{-4}	JACS.1954.76.3820.X
131	XeF_8	Sc.1971.173.1238.X
135	$Ce(NO_3)_6^{-3}$	JCP.1963.39.2881.X
143	B_2H_6	JACS.1951.73.1482.E
145	$FE_2(CO)_9$	JCS.1939.–.286.X
149	P_4	JCP.1935.3.699.E
151	$B_{12}H_{12}^{-2}$	JACS.1960.82.4427.X
155	C_{60}	Sc.1991.254.410.E
161	$(SiO_2)_x$	JACS.1925.47.2876.X
177	H_2	CJR.1950.A,28.144.I
177	H_2^+	Herzberg.1950.UV
177	He_2^+	Herzberg.1950.UV
177	LiH	Herzberg.1950.UV
177	NaH	Herzberg.1950.UV
177	KH	Herzberg.1950.UV
177	HF	Herzberg.1950.I
177	HCl	Herzberg.1950.I
177	HBr	SA.1952.5.313.I
177	Li_2	Herzberg.1950.UV
177	Na_2	Herzberg.1950.UV
177	K_2	Herzberg.1950.UV
177	N_2	CJP.1954.32.630.R
177	N_2^+	Herzberg.1950.UV
177	O_2^{-2}	AC.1954.7.838.X
177	O_2^-	AC.1955.8.503.X
177	O_2	JCP.1955.23.1739.E
177	O_2^+	Herzberg.1950.UV
177	F_2	CJP.1951.29.151.R
177	Cl_2	Herzberg.1950.UV
177	Br_2	JCP.1955.23.1739.E
177	CN^-	RTC.1942.61.561.X
177	CN	AJ.1954.199.303.UV
177	CN^+	Herzberg.1950.UV
177	CO	JCP.1949.17.1099.I
177	CO^+	Herzberg.1950.UV
177	NO	JCP.1955.23.57.I
177	NO^+	CJP.1955.33.355.UV
179	HCCH	AC.1950.3.46.E
179	HCN	JCP.1953.21.448.I
179	CH_2CCH_2	CJP.1955.33.811.R
179	CH_2CO	JCP.1953.21.1898.I+M
179	HNCO	JCP.1950.18.990.I+M
179	CO_2	JOSA.1953.43.1037.I
179	NO_2^+	AC.1950.3.290.X
179	CH_2N_2	SAWW.1935.144.1.E
179	HN_3	JCP.1950.18.1422.M
179	N_2O	JCP.1954.22.275.I+M
179	HNCS	JCP.1953.21.1416.M
179	COS	JCP.1935.3.821.E
179	CS_2	ACS.1947.1.149.E
179	XeF_2	JACS.1963.85.241.N
181	HNCO	JCP.1950.18.990.I+M
181	HN_3	JCP.1950.18.1422.M
181	HONO	JCP.1951.19.1599.I
181	FNO	JCP.1951.19.1071.M
181	ClNO	PR.1951.83.431.M
181	BrNO	JACS.1937.59.2629.E
181	NO_2^-	AC.1955.8.852.X
181	NO_2	JOSA.1953.43.1045.I
181	O_3	JCP.1953.21.851.M
181	SO_2	JCP.1954.22.904.M
183	H_2O	PR.1954.95.374.M
183	OF_2	JPC.1953.57.699.E
183	CH_3OCH_3	JACS.1935.57.473.E
		& JACS.1935.57.2684.E
183	OCl_2	JCS(A).1966.–.336.M
183	H_2S	PR.1954.94.1203.M
183	F_2S	Sc.1969.164.950.M
183	CH_3OCH_3	JCP.1961.35.479.M
183	SCl_2	JACS.1938.60.2872.E
185	BF_3	JACS.1937.59.2085.E
185	CH_2CH_2	JCP.1942.10.88.I
185	CH_2CCH_2	CJP.1955.33.811.R
185	CH_2CO	JCP.1953.21.1898.I+M
185	CH_2O	JCP.1954.22.289.I+UV
185	CHFO	JCP.1961.34.1847.M
185	CF_2O	JCP.1962.37.2995.M
187	HCO_2H	JCP.1955.23.210.M
187	CCl_2O	JCP.1953.21.1741.M
187	CO_3^{-2}	JACS.1937.59.1380.X
187	NO_3^-	AC.1950.3.290.X
187	NO_2F	JCS(A).1968.–.1736.M
187	SO_3	JACS.1938.60.2360.E
189	ClF_3	JCP.1953.21.609.M
189	BrF_3	JCP.1957.27.223.M
189	XeF_4	Sc.1963.139.1208.X

Index